PSYCHOLOGICAL ISSUES

VOL. VIII, No.4 MONOGRAPH 32

SCIENTIFIC THOUGHT AND SOCIAL REALITY: ESSAYS BY MICHAEL POLANYI

Edited by
FRED SCHWARTZ

INTERNATIONAL UNIVERSITIES PRESS, INC.
239 Park Avenue South • New York, N.Y. 10003

Library of Congress Cataloging in Publication Data

Polanyi, Michael, 1891-
Scientific thought and social reality; essays.

(Psychological issues, v. 8, no. 4. Monograph 32)
Bibliography: p.
1. Science–Philosophy. 2. Knowledge, Theory of.
I. Title. II. Series.
Q175.P822 501 74-5420
ISBN 0-8236-6005-2

Manufactured in the United States of America

PSYCHOLOGICAL ISSUES

Subscription per Volume, $20.00

Single Copies of This Number, $6.50

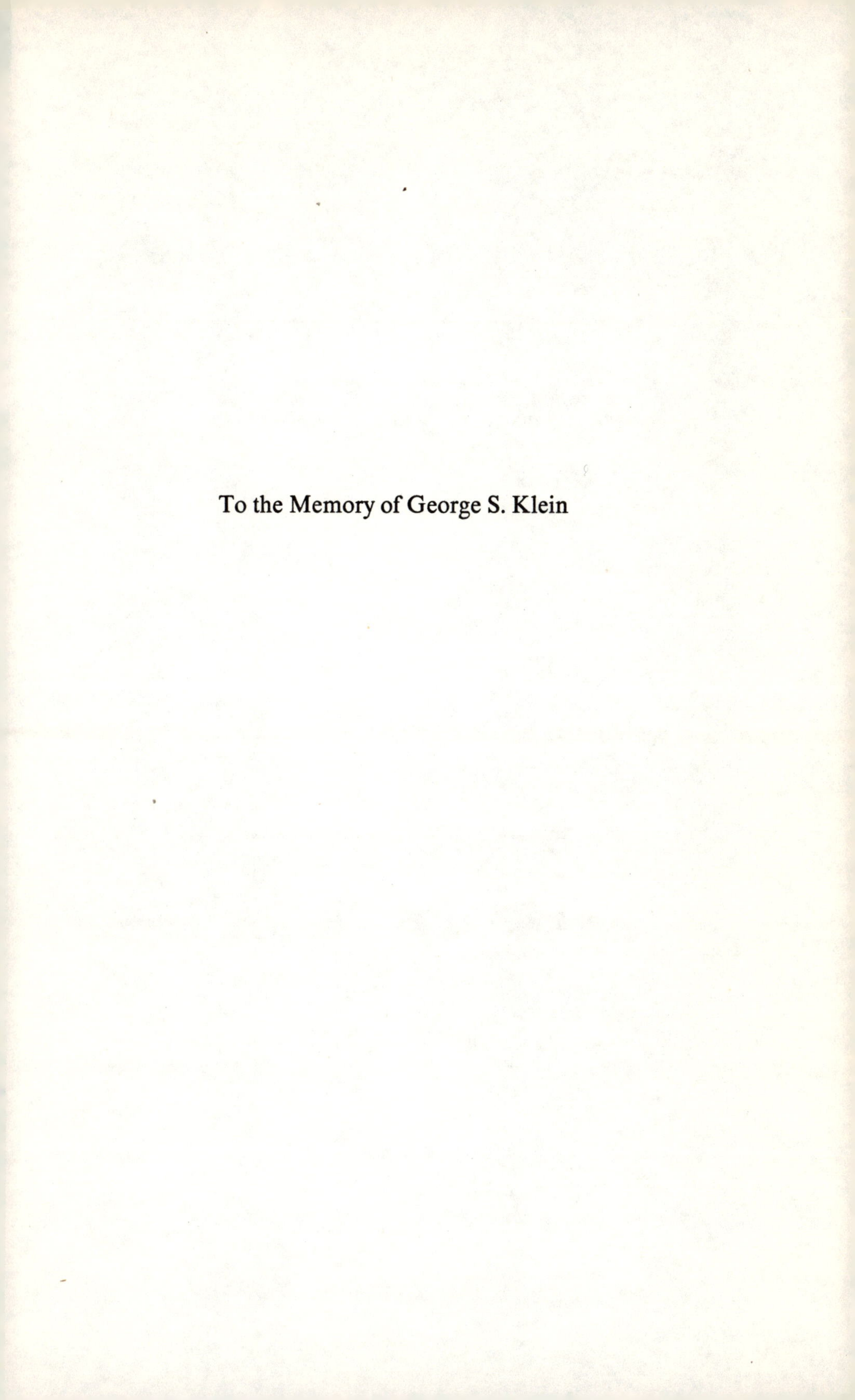

To the Memory of George S. Klein

CONTENTS

ACKNOWLEDGMENTS

Acknowledgment is gratefully made to the following publishers and periodicals for permission to reprint the essays in this collection:

Science (Chapter 1), The Manchester Literary and Philosophical Society (Chapter 2), The British Association for the Advancement of Science (Chapter 3), Twentieth Century Magazine Ltd. (Chapter 4), The University of Chicago (Chapters 5 and 8), *Cambridge Journal* and Bowes & Bowes (Chapter 6), *Encounter* (Chapters 7 and 9).

FOREWORD

MICHAEL POLANYI

My meetings with George Klein aimed mainly at tracing thought processes shaped by immediate experience. I was looking for these with a view to the structure of scientific discoveries. I was made happy by the response I received from Klein; I hardly expected it from a scientist engaged in other inquiries. It was a happy event; but the papers republished here had different grounds.

The articles printed here were not exactly strange to George Klein's work, but their scope was different from it. They dealt with man's mental position within sciences aiming at mechanical explanations. I brought with me a general interest that I shared with some other scientists and philosophers. Klein joined us in this matter very warmly. This is how he came to collect a set of my articles for republication in the journal he was directing; his colleagues have told me about this. I was moved by this link to George, whom I loved and admired.

I would add a word here: the reader may notice some repetitions, which I hope he will set aside as incidental: they are parts of a battle I shared with George Klein. And I may also excuse them, since the troubles they spoke of seem more urgent today than they were at that time.

PREFACE

FRED SCHWARTZ

George Klein spent the 1967-1968 academic year on sabbatical at the Austin Riggs Center. That was my last year at Riggs and I was fortunate to have an office adjoining his. I often met with Klein to discuss some specific issue, and the discussion invariably swept on to other topics, reflecting his broad range of interests, incisive intellect, and refusal to be pinned down to any one question. The name Michael Polanyi came up again and again in these talks. Klein had me read Polanyi's book, *Personal Knowledge,* and lent me a number of Polanyi's essays. He believed that Polanyi's work provided a philosophical framework for a psychology of man, and in particular for psychoanalysis. At the time, he was working on a book that drew upon his recently completed psychoanalytic training, prior clinical experience and research, and Polanyi's writings. I remember thinking that Klein was at a crossroads in his career, in part because of Polanyi's influence.

At some point in this stimulating year, Klein asked me to collect Polanyi's essays on scientific method. He thought that Polanyi's general contribution was well represented in his books, but that his specific contribution to scientific method was largely unknown to psychologists and psychoanalysts. This seemed like a modest enough task, until I found myself reading 30 to 40 essays and trying to select from these a collection for publication in a single monograph. Klein's enthusiasm, however, was not misguided—I found all the essays stimulating and provocative.

There were delays in collecting the essays, and further delays in meeting with Klein after we both left Riggs. The project was set aside at the time of Klein's death, as I had never really established how he wanted this material organized and presented.

In the years that followed my introduction to Polanyi's writings, there has been a gradual reduction in funding for basic science and research training. What began as a realistic pruning of excesses in the funding of science gradually evolved into an attack on science itself. In some alarm, I wrote Herbert Schlesinger in May, 1972, and asked that the Polanyi essays be considered for publication in *Psychological Issues* because of their direct bearing on what was happening at the time. There followed correspondence with Polanyi and Schlesinger, both of whom supported the project and gave generously of their time. As of this writing, the essays appear to be even more topical than they were in 1972. In a recent letter, Polanyi wrote, "I am profoundly concerned with the widespread attempts at the destruction of freedom in science." Those are strong words, but do not seem to overstate the issue.

Polanyi and I are at one accord in dedicating the publication of these essays to the memory of George Klein.

INTRODUCTION

FRED SCHWARTZ

How does the scientist think? How does his thought influence society? How should science be administered so as to facilitate scientific work? These are the kinds of questions that have concerned Michael Polanyi for the past 30 years. At the time that he first wrote about such matters, there seemed little doubt about the status of science in society. Most people believed that science provided a meaningful description of reality and valued the scientist's commitment to reason, evidence, and rational discourse. Science was viewed as a dominant force in shaping our material existence and personal beliefs. It was even believed that the scientific attitude fostered the development of a free society. At one time, Polanyi could thus assert that "In the great struggle for our civilization science occupies a section in the front line" (this volume, Chapter 3), and arouse little opposition.

Today, science is under fire. It has become an increasingly expensive form of human activity and does not always have benign consequences. Serious questions have therefore been raised concerning the scientist's traditional autonomy to pursue knowledge for its own sake. But any limitation on the scientist's autonomy threatens the value position of free inquiry that is central to scientific work. We are thus faced with a dilemma that requires thoughtful deliberation. Instead, the issue has become polarized along ideological and political lines. Changes in funding patterns and restrictions on areas of investigation are being dictated by conflicting social and economic pressures. The scientist reflects these conflicting trends in his search for financial security, academic status, and the opportunity to follow his own interests and hunches. Is our scientific heritage being eroded? Will we continue to value the lonely pursuit of knowledge by a few gifted men? Only time will tell. Some among us, however, are not sanguine about the future. We may be witnessing a transient phase in the history of science or its destruction as a social institution.

Polanyi began his professional career as a physical chemist in 1917 and continued to work as an exact scientist until 1948. This was a period of major theoretical development in physics and allied sciences. A new view of reality was developed because of the highly creative thought of a handful of men. This same period, however, gave rise to a utilitarian outlook toward science. Polanyi perceived this latter development as a threat to how he worked as a scientist and so retired from scientific endeavors to write about how the scientist actually thinks and works. He was thus one of the first to realize that the relationship between science and society was undergoing a significant and painful change.

The present collection of essays has three main themes: a critique of logical positivism; a cognitive theory of epistemology; and a discussion of the scientist's need for autonomy in pursuing knowledge. Polanyi is throughout a champion of the scientist's creative urge to uncover the truth. In the remainder of this introduction, I would like to discuss some background issues in the psychology of scientific thought, consider Polanyi's position in the light of recent developments in psychology, and then briefly return to the question of the need for autonomy in scientific investigation.

Modern science began with the radical idea that knowledge can be gained through observation and experimentation. That this was indeed a novel idea is evident if we consider the amalgamation of religious and Aristotelian thought in the Middle Ages, with its emphasis on teleology, qualities of experience, and the central position of man in the scheme of things (Burtt, 1924). From the beginning of the modern era, there have been two main views of how we formulate ideas about reality. Francis Bacon promulgated the inductive method, by which one reasons from specifics to general hypotheses. Descartes argued that one can intuit hypotheses directly. In either case, one has a hypothesis in hand that one did not have before. One can then make deductions and test them by experimentation.

The above distinction breaks down in actual practice, as both methods are employed simultaneously. There are always some facts underlying a new hypothesis, and there is always an inductive leap from the parts to the whole. Differentiation and integration are basic parameters of human thought. Nevertheless, there is a tendency to view induction as consistent with the empirical approach and to view the Cartesian method as outside the scope of science. There seems to be less mystery about going from specifics to general conclusions, and so it seems more reasonable to believe that we will eventually be able to specify how this comes about. This belief is reflected in the recent development of computer programs for induction. It has even been said that men are not above the laws of logic "that govern every automaton's behavior" (Huesmann and Cheng, 1973).

Induction and hypothesis formation are not only scientific methods but also primitive theories of thinking. As methods, they describe the way in which one develops new ideas. As theories of thinking, they refer to how this development is accomplished. One can even argue that the inductive and Cartesian methods are modes of behavior that operationally define how the scientist thinks. So all one need do is study how scientists arrive at their ideas. The difficulty with this approach is that scientists rationalize their thought so as to achieve respectability, or are so committed to empiricism as to deny their own creativity. Newton, for example, claimed that he "deduced" his laws from observation and denied the legitimacy of hypothesis formation, although many of his ideas were highly original and even speculative. Evidence of original scientific thought includes many of the great discoveries in physics. In his dialogue with Wertheimer (Wertheimer, 1959), Einstein acknowledged that the axioms in the Einstein-Infeld book did not represent his thought. "No really productive man thinks in such a paper fashion" (quoted by Wertheimer, 1959, p. 228). Gamow's (1966) account of quantum theory and quantum mechanics provides many examples of highly original thought based on minimal data. The original theories in these fields seemed bizarre and startling even to their originators. But the classic example is Copernicus, who initiated a revolution in human thought by going against everyday experience, because of his conviction that nature is governed by simple mathematical laws. It is difficult to imagine how such a conviction could be derived from sense experience.

The inductive method has been the basis of major advances in the natural sciences. But the method itself is the source of two difficulties. The first is based on Hume's dictum that no theory can be proven inductively. There is always a heuristic gap, to use Polanyi's term, between observation and theory. We tend to overlook this gap because the ability to abstract is taken for granted. Newton misjudged his own genius when he said that he deduced laws from observation–his contribution was a major intellectual feat.

The second difficulty goes back to the inception of the atomic theory of matter. According to the Greek philosopher Democritus, "Sweetness exists only by convention, bitterness by convention, color by convention; the atoms and the void alone exist in reality." In this view, sweetness and color are "secondary" or mental qualities, not attributes of real physical objects. The world of everyday experience is therefore an illusion. The atoms and the void, however, cannot be directly observed. Thus we seem to be left with a physical world that cannot be directly apprehended by mind and a world of subjective experience that cannot be explained in the language of matter theory. How, then, can the scientist gain knowledge of reality? Man looked into the abyss and pulled back, asserting that the origin of knowledge is direct experience. In this view, atoms in motion create a moving picture in the mind. All the scientist can do is describe this panorama as it unfolds. But if the mind can somehow fabricate sense experience out of matter in motion, why can't it also fabricate knowledge? The former is as much an achievement as the latter.

Ryle (1962) has cogently argued that this dilemma is mostly semantic. In essence, the scientist and everyman are not talking about the same thing at all, so one point of view does not invalidate the other. But it must be kept in mind that the inductive method has its roots in this dilemma. It is a method based on sense experience. The scientist formulates mathematical theories about the physical universe on the basis of observations that belong, in principle, to everyman's world of subjective experience. He converts the "quality" of experience into mathematical quantities that are "real," in the sense that they refer to properties or relationships in the physical world. In different forms, this is the mind-body problem, the problem of epistemology, and the problem of creative thought. The inductive method is a beautiful method for acquiring knowledge about reality, but it may be an inexact theory of mind.

We nevertheless cannot decide whether the scientist can intuit hypotheses directly by looking at the end result of scientific thought. The scientist who claims originality may do so because he cannot specify all the bits and pieces that contributed to his thought, or does not wish to do so because his thought, while inductive, was not entirely rigorous. Furthermore, his claims may be a reaction to long periods of frustration and gloom, now overthrown at one blow. One can well excuse grandiose or manic indulgence at the moment of discovery. Kepler's outburst (quoted by Polanyi in Chapter 7) on discovering his laws of planetary motion is such an example; it is more a clue to his frame of mind than to his thought.

Let us turn, then, from how the scientist works, or appears to work, to the psychology of thought. Polanyi has taken both tacks in his essays and books, giving numerous examples of what scientists actually do, but also developing a theory of the process of scientific thought.

Theories of thought can be divided into two broad categories—additive theories, in which the elements (ideas, traces, etc.) are linked together by some kind of bond, and synthetic theories, in which the thought process is attributed to properties of the system. This distinction is more easily approached in the field of perception, where the same dichotomy is found.

Some of the relevant data have been reviewed by Held (1965), Pribram (1971), and Neisser (1967). Additive theories assume some form of direct correspondence between the physical object and its mental representation (Held, 1965). For example, it was once thought that objects give off copies (eidola) that directly enter the eye. We now know that the visual image is projected onto a screen behind the lens, that this image is inverted and reversed, that the lens distorts the image, and that information from the eye is transmitted as a series of nerve impulses to the visual cortex. We also know that the percept remains more or less constant in size, shape, and color, although the object is seen at different distances, at different angles, and in different illumination. Furthermore, many "objects" are markedly variable in form and yet are rarely misperceived–for example, patterns of speech and handwriting vary across people and across time, and yet are easily recognized. Object constancy is thus maintained in the face of manifold distortions and transformations. Facts of this nature led the above authors to conclude that the percept is not built up from elements (including feature analyzers in the brain) but is, instead, the product of a "constructive" or synthetic process. Cognitive theorists (e.g., Neisser) leave open how the nervous system achieves this construction. Others are developing neurological models to handle these facts. Pribram (1971), for example, has suggested that a "neural holographic" process may underlie this synthesis. Such theories are admittedly speculative, but the trend is toward postulating and discovering structural mechanisms in the brain to account for the richness and complexity of perceptual experience.

Returning to the psychology of thought: again there are theories of correspondence. In the classical view, ideas are faint copies of visual images–copies of copies, linked together by associative bonds. The behaviorist speaks instead of responses linked to stimuli. But what the response seems to preserve is meaning rather than the quality of the stimulus. Thus, the same "response" can be initiated by any limb, or by the same limb in different starting positions. At a more molar level, Bartlett's (1932) classic study suggests that remembering is an active, constructive process, not a passive reduplication of past events. There seem to be systems that can construct memories and thoughts out of previous constructions–holograms of holograms, as it were.

Polanyi's theory of knowledge is essentially cognitive. On the basis of his own experience, he concluded that the scientist does not build theory by adding up facts–he is capable, instead, of constructing a new view of reality. As we have seen, however, one can argue that the scientist's belief in his own constructions is an illusion, albeit useful as strategy. According to Reichenbach (1953), "The philosopher of science is not much interested in the thought processes which lead to scientific discoveries... he is not interested in the context of discovery, but in the context of justification."

Reichenbach then adds this significant sentence: "But the critical attitude may make a man incapable of discovery; and, as long as he is successful, the creative physicist may very well prefer his creed to the logic of the analytic philosopher" (p. 197). So it may be useful to believe in one's hunches. Such belief carries one through long periods of arduous and frustrating work and sustains curiosity. Nevertheless, the scientist's motives for doing research and his beliefs, hunches, and intuition are extrascientific baggage, without formal methodological status. The only reality is that which is directly observed. As we have tried to show, the attempt to exclude originality from science in the name of empiricism requires that we ignore certain observations. We may do so in the belief that all thought can eventually be explained in terms of an additive model, but this belief represents an intellectual bias and is itself not verifiable. The belief in original thought is also a bias. In the last analysis, we must choose between these two points of view on the basis of their utility.

In psychoanalytic writings, creative thought is described as biphasic. There is an "inventive" phase that follows the rules of the primary process, and an "elaborative" phase that is governed by secondary-process rules of logic (Rapaport, 1951). The inventive phase belongs to the "context of discovery" and includes the ability to see new relationships. The scientist understandably values secondary-process thought because it is only such thought that leads to proof. Within the psychoanalytic framework, inventive thought is, by definition, irrational and unpredictable. It may nevertheless be central to scientific work. Training scientists to avoid speculation may make for sterile science and impede the acquisition of knowledge. As psychologists, we would argue that the danger of a meaningless and boring science is greater than the danger of the irrational.

How the scientist thinks cannot be divorced from his social reality. Western science first appeared in modern form in the thirteenth century, but it was not until 400 years later that changes in society made a true development of science possible. Science, in turn, has profoundly influenced daily life. Such changes, however, did not come about without opposition. Galileo had to renounce his belief in the Copernican system before the Inquisition and Giordano Bruno was burnt alive for his beliefs. In the eighteenth century, scientific thought was attacked by both the Church and the humanists. The scientific movement nevertheless took hold and dominated man's thought in the next century. The Romantics could argue that there was more to man's nature than blind mechanism, but no serious thinker really questioned the facts of science or its method. It became the scientist's task to seek the truth and society's task to live with the consequences.

It is part of our tradition that scientists establish their own norms and judge each other's work according to a system of peer review. Within this system of constraints, they pursue their own interests and ideas, under the influence of a variety of motives and needs, from fame, power, financial security, and the wish to please others, to curiosity and a desire to learn the truth. Such a hodgepodge of social, emotional, and intellectual forces may seem to make for a fragile motivational structure, and yet it is precisely this kind of structure that has led to the scientific achievements of the past 300 years and thus to the shaping of life as we know it today. There are pressures today to change all this. The movement is toward training fewer investigators, contract research, support of socially relevant research, and the organization of researchers into groups so that they can coordinate their work when dealing with the same problem. Such pressures are especially felt in biomedical research because the government funded most of such work in the 25-year period following World War II. Basic science is most vulnerable at this time as economic, social, and political considerations combine to shift funds away from pure research to improving our daily lives. There is also growing pressure to restrict investigations that directly bear on the public. The recent controversy over intelligence testing is a case in point. Whether or not these developments are good for society is not the issue here. The point is simply that for either the right or the wrong reasons, the scientist has lost some of his traditional autonomy. The impact of science on daily life would seem to call for some degree of accountability–witness the effects of atomic research in the early part of this century on our lives today–and yet accountability would seem to threaten the structure of science as it has evolved over the years. The question is, quite simply, can we have accountability and autonomy, or are they mutually exclusive?

As stated, this question easily leads to polarization. One is either for the truth or for the good of man. And yet we recognize that the pursuit of knowledge has contributed to the general good. How this issue is resolved will depend upon the interplay of forces within and outside of science. In working out a solution, we have to avoid clichés and simple answers. It is here the Polanyi's wisdom is most helpful. What he has shown is that there is an inner connection between scientific thinking and how scientific activity is supported by society. He has tried to show again and again that scientific discovery requires an atmosphere of free inquiry, constrained only by the scientist's own sense of truth and the opinion of his scientific peers. This fact must be taken into account in deciding how science is to be funded. Matters would be greatly simplified if science were merely a collection of facts, based essentially on the inductive method. If that were true, then one could simply order research, based on social need, with little regard to the scientist's own creativity and motivation. But if it is not true, then some allowance must be made for the ability of scientists to formulate new ideas, based on their own good sense. It seems clear that no funding agency would have had the foresight to order Kepler to work out the laws of planetary motion, Einstein to develop the special theory of relativity, or in our own time, Watson and Crick to formulate the chemical structure of DNA. To quote Watson, their historical achievement was "an adventure characterized by youthful arrogance and by the belief that the truth, once found, would be simple as well as pretty" (1968, p. ix). Arrogance, belief, simplicity, and beauty–these are the ingredients that Polanyi found in the writings of many great scientists. These are irrational ingredients, not readily computerized or used as justification for spending public funds.

A final way of evaluating Polanyi's contribution is in terms of the concept of splitting, as employed by the English analytic school, although I hasten to add that Polanyi would probably not endorse this terminology. Nevertheless, what he has fought against all these years is the splitting off of scientific activity from the scientist's motives and his belief in his own rational powers to construct a meaningful picture of reality. In a psychological sense, splitting dehumanizes people and alienates them from society. In scientific work, splitting results in the depersonalization of the scientist and the derealization of his findings. There is no excitement, no discovery, no creative tension or sense of accomplishment. There is no "intellectual passion," to use Polanyi's evocative term. The end result is often banal research and a caricature of the inductive method, in which science indeed becomes a mere accumulation of facts. But what is even more troublesome, splitting results in a pseudo objectivism, because the scientist's motives are denied, in the name of objectivity, only to enter into his choice of work by the back door, from a scientific id. One thing we have learned from psychoanalysis is that motives cannot be incorporated into our reality unless they are acknowledged. From the vantage point of both a scientist and a philosopher, Polanyi has tried to show that science is basically a human activity.

1

THE AUTONOMY OF SCIENCE

Today the position of science, which was unquestionably accepted in the Western countries for the last 300 years or so, has been challenged by an authoritarian doctrine.

It is difficult to trace a complete authoritative statement of the argument used in support of the state control of science. But I believe that in its most precise form this argument would run about as follows. "No scientific statement is absolutely valid, for there are always some underlying assumptions present the acceptance of which represents an arbitrary act of faith. Arbitrariness prevails once more when scientists choose to pursue research in any one direction rather than another. Since the contents of science and the progress of science both vitally concern the community as a whole, it is wrong to allow decisions affecting them to be taken by private individuals. Decisions such as these should be reserved to the public authorities who are responsible for the public good. It follows that both the teaching of science and the conduct of research must be controlled by the state."

I believe this reasoning to be fallacious and its conclusions to be wrong. Yet I shall not try to meet the argument point by point, but will instead oppose it as a whole by analyzing the actual state of affairs which it profoundly misrepresents. I shall survey the individuals and groups who normally make the decisions which contribute to the growth and dissemination of science. I shall show that the individual scientist, the body of scientists, and the general

First published in *The Scientific Monthly*, 60 (Feb.):141-150, 1945.

public each play their part and that this distribution of function is inherent in the process of scientific development, so that none of these functions can be delegated to a superior authority. I shall argue that any attempt to do this could only result in the distortion and—if persisted in—in the complete destruction of science. I shall demonstrate instances where such attempts have actually been made and the destruction did actually come to pass.

The primary decisions in the shaping of scientific progress are made by individual investigators when they embark on a particular line of inquiry. Today in science the independent investigator is usually a professional scientist, appointed by public authorities in view of his scientific record to a post where he is expected to do research. For this he is given freedom to use his own time and is often also given control over considerable means in money and personnel.

The granting of such discretion to individuals for the purposes of their profession is fairly common in all departments of life. Holders of higher posts in business, politics, the law, medicine, the army, the church, are all invested with powers which enable them to follow their own intuitive judgment within the framework of certain rules. They use this freedom in order to discharge their duties. Yet the degree of independence granted to the scientist may appear to be greater than that allowed to other professional men. A businessman's duty is to make profits, a judge's to find the law, a general's to defeat the enemy; while in each case the choice of the specific means for fulfilling the task is left to the judgment of the person in charge, yet the standards of success are laid down for him from outside. For the scientist this may not hold to quite the same extent. It is part of his commission to revise and renew by pioneer achievements the very standards by which his work is to be judged. He may be denied full recognition for a considerable time—and yet his claims may be ultimately vindicated. But the difference is only one of degree. All standards of professional success undergo some change in the course of professional practice, and on the other hand even the most daring pioneer in science accepts the general conceptions of scientific achievement and bases his scientific claims essentially on traditional standards. I shall have more to say in this connection later.

In any case, the power to use his own intuitive judgment and the encouragement to embark on original lines of inquiry are not given to the scientist to enable him to exercise his own personal wishes. The high degree of independence that he enjoys is granted only to enable him to discharge the more effectively his professional obligations. His task is to discover the opportunities in the given state of science for the most successful application of his own talents and to devote himself to the exploitation of these openings. The wider his freedom, the more fully can be throw the force of his personal conviction into the attack on his own problem.

At the start his task is hidden, but it is none the less definite. There is ample evidence to show that at any particular moment the next possibilities of discovery in science are few. The next step to be taken in any particular field is in fact sometimes so clear that we read of a "dramatic race" between leading scientists for an impending discovery. A series of such races took place within a period of a few years for the discovery of the synthesis of various vitamins. In 1935 Karrer in Zurich and Kuhn in Heidelberg competed in the synthesis of vitamin B_2. In 1936 three teams, Andersag and Westphal in Germany, Williams and Cline in the United States, and Todd and Bergel of England raced for the synthesis of vitamin B_1. And in 1938 one of the participants in the B_1 race, Todd, and one in the B_2 race, Karrer, rivalled closely in the synthesis of vitamin E. Only a few years earlier (1930) a great race was won in physics when Cockcroft and Walton, working under Rutherford's guidance in Cambridge, accomplished the artificial disintegration of the atom by electric discharge–ahead of Lange and Brasch in Germany and Breit, Tuve, Hafstad, Lauritsen, Lawrence, and others in America. Or to take an example in pure theoretical physics: between 1920 and 1925 the standing problem of theoretical physicists was the reconciliation of classical mechanics and quantum theory; and around the year 1925 a number of physicists (de Broglie, Heisenberg, Born, Schrödinger, Dirac) did actually discover–more or less independently–the various parts of the solution. In a review of Eve's biography of Rutherford, Sir Charles Darwin estimates roughly by how much Rutherford may have anticipated his contemporaries with his various discoveries, and suggests for most cases spans of time ranging from a few months to three or four years. Rutherford himself is quoted as saying that no one can see more than an eighth of an inch beyond his nose and that only a great man can look even as far as that.

Scientific research is not less creative and not less independent because at any particular time only a few discoveries are possible. We do not think less of the genius of Columbus because there was only one New World on this planet for him to discover.

Though the task is definite enough, the solution is none the less intuitive. It is essential to start in science with the right guess about the direction of further progress. The whole career of a scientist usually remains linked to the development of the single subject which stimulated the early guess. All the time the scientist is constantly collecting, developing, and revising a set of half-conscious surmises, an assortment of private clues, which are his confidential guides to the mastery of his subject.

This loose system of intuitions cannot be formulated in definite terms. It represents a personal outlook which can be transmitted only—and only very imperfectly—to personal collaborators who can watch its daily application for a year or two to the current problems of the laboratory. This outlook is as much emotional as it is intellectual. The expectations which it entertains are not mere idle guesses but active hopes filled with enthusiasm.

The emotions of the scientist also express and uphold the values guiding research; they turn with admiration to courage and reliability and pour scorn on the commonplace and the fanciful. Such emotions again can be transmitted only by direct contact in the course of active collaboration. They are in fact the very lifeblood of collaboration in a research school. Its leader has no more important function than to maintain enthusiasm for research among his students and instill in them the love of his own particular field.

Such is the calling of the scientist. The state of knowledge and the existing standards of science define the range within which he must find his task. He has to guess in which field and to what new problem his own special gifts can be most fruitfully applied. At this stage his gifts are still undisclosed, the problem is yet obscure. There is in him a hidden key capable of opening a hidden lock. There is only one force which can reveal both key and lock and bring the two together: the creative urge which is inherent in the faculties of man and which guides them instinctively to the opportunities for their manifestations. The world outside can help by teaching, encouragement, and criticism, but all the essential decisions leading to discovery remain personal and intuitive. No one with the least experience of a higher art or of any function requiring higher judgment could conceive it to be possible that decisions such as these could be taken by one person for another. Decisions of this kind can in fact only be suppressed by the attempt to transfer them to an outside authority.

The scientist today cannot practice his calling in isolation. He must occupy a definite position within a framework of institutions. A chemist becomes a member of the chemical profession; and a zoologist, a mathematician, or a psychologist each belongs to a particular group of specialized scientists. The different groups of scientists together form the scientific community.

The opinion of this community exercises a profound influence on the course of every individual investigation. Broadly speaking, while the choice of subjects and the actual conduct of research is entirely the responsibility of the individual scientists, the recognition of claims to discoveries is under the jurisdiction of scientific opinion expressed by scientists as a body. Scientific opinion exercises its power largely informally but partly also by the use of an organized machinery. At any particular time only a certain range of subjects is deemed by this opinion to be profitable for scientific work. Accordingly no training is given and no posts either for teaching or for research are offered outside these fields, while existing research schools are specialized in these subjects and so are the journals available for publication.

Even within the fields that are recognized in this sense at any particular time, scientific papers cannot be published without preliminary approval by two or three independent referees, called in as advisers by the editor of the journal. The referees express an opinion particularly on two points: whether the claims of the paper are sufficiently well substantiated, and whether it possesses a sufficient degree of scientific interest to be worth publishing. Both characteristics are assessed by conventional standards which in fact are changed from time to time according to variations of scientific opinion. Sometimes it may be felt that the tendency among authors is toward too much speculation, which the referees will then try to correct by imposing more discipline. At other times there may seem to be a danger of absorption in mere mechanical work, which referees will again try to curb by insisting that papers should show more originality. Naturally, at different periods there are also marked variations as regards the conclusions that are considered sufficiently plausible. A few years ago there was a period in which it was easy to get a paper printed claiming the transformation of chemical elements by ordinary laboratory processes; today—as in earlier times—this would be difficult, if not altogether impossible.

The referees advising scientific journals may also encourage those lines of research which they consider to be particularly promising, while discouraging other lines of which they have a low opinion. The dominant powers in this respect are, however, exercised by referees advising on scientific appointments, on the allocation of special subsidies, and on the award of distinctions. Advice on these points, which often involve major issues of the policy of science, is usually asked from and tendered by a small number of senior scientists who are universally recognized as being the most eminent in a particular branch. They are the chief Influentials, the unofficial governors of the scientific community. By their advice they can either delay or accelerate the growth of a new line of research. New facilities for work can be most rapidly made available by the granting at their command of special subsidies for research. By the award of prizes and of other distinctions they can invest a promising pioneer almost overnight with a position of authority and independence. More slowly, but no less effectively, a new development can be stimulated by the policy pursued by the Influentials in advising on new appointments. Within 10 years or so a new line of thought may be established by the selection of appropriate candidates for chairs which have fallen vacant during that period. The same end can be promoted by the setting up of new chairs, which sometimes replace others which have become obsolete.

The constant redirection of scientific interest by the leaders of scientific opinion fulfills the important function of keeping the standards of performance in different branches of science at an approximately equal level. In the various branches the standards of reliability and systematic interest are applied in somewhat different ways. In general, the greater the human interest of the subject matter, the less rigorous the tests required for establishing the same standard. Living beings are intrinsically more interesting than inanimate nature. Scientific statements will be allowed to be less definite and less certain if made about plants or animals than about minerals or stars. Similarly, a speculative achievement of modest range may be recognized as a success if it relates to the problems of living matter. The leaders of scientific opinion have to adjust the different standards in such a manner as to maintain in every field a uniform level of development–this level being jointly characterized by the intrinsic interest of the subject matter, the profundity or systematic interest of the generalizations involved, and the precision and certainty of the new statements made.

The steady equalization of standards in all branches is necessary, not only in order to maintain a rational distribution of resources and recruits for research schools throughout the field of science, but also in order to uphold equally in every branch the authority of science with regard to the general public. With the relation of science and the public I shall shortly deal in some detail. But a particular aspect of it requires mention at this point since it involves the final phase of the process by which recognition is given to new scientific claims. Published papers are open to discussion and their results may remain controversial for some time. But scientific controversies are usually settled–or else shelved to await further evidence–within a reasonable time. The results then pass over into textbooks for universities and schools and become part of generally accepted opinion. We note that this final process of codification is again under the control of the body of the scientific opinion–expressed by reviewers–under whose authority textbooks are in fact brought into circulation.

The standards of science–like those of all other arts and professions–are transmitted largely by tradition. Science in the modern sense originated some 300 years ago from the work of a small number of pioneers, among whom Vesalius and Galileo, Boyle, Harvey, and Newton were pre-eminent. The founders of modern science discussed extensively and with considerable insight the new methods which they applied; moreover, the doctrines of the contemporary philosophy–particularly through John Locke–gave full expression to their outlook. Yet the core of the scientific method lies in the practical example of its works. Whatever the various philosophies of the scientific method may still reveal, modern science must continue to be defined as the search for truth on the lines set by the examples of Galileo and his contemporaries. No pioneer of science, however revolutionary—neither Pasteur, Darwin, Freud, nor Einstein—has denied the validity of that tradition or even relaxed it in the least. The great succession of men of genius to whose creative powers science has given scope since the end of the sixteenth century has not overshadowed the first pioneers but has, on the contrary, increasingly revealed the implications of their discoveries and thus added ever more brilliance to their achievements.

Modern science is a local tradition and is not easily transmitted from one place to another. Countries such as Australia, New Zealand, South Africa, Argentina, Brazil, Egypt, Mexico, have built great modern cities with spacious universities, but they have rarely succeeded in founding important schools of research. The total current scientific production of these countries before the war was still less than the single contributions of Denmark, Sweden, or Holland. Those who have visited the parts of the world where scientific life is just beginning know of the backbreaking struggle that the lack of scientific tradition imposes on the pioneers. Here research work stagnates for lack of stimulus, there it runs wild in the absence of any proper directive influence. Unsound reputations grow like mushrooms, based on nothing but commonplace achievements or even on mere empty boasts. Politics and business play havoc with appointments and the granting of subsidies for research. However rich the fund of local genius may be, such an environment will fail to bring it to fruition. In the early phase in question New Zealand loses its Rutherford, Australia its Alexander and its Bragg, and such losses further retard the growth of science in a new country. Rarely, if ever, was the final acclimatization of science outside Europe achieved until the government of the overseas country succeeded in inducing a few scientists belonging to some traditional center to settle down in their territory and allowed the newcomers to develop there a new home of scientific life moulded by their own standards. This demonstrates perhaps most vividly the fact that science as a whole is based—in the same way as the practice of any single research school—on a local tradition, consisting of a fund of intuitive approaches and emotional values which can be transmitted from one generation to the other only through the medium of personal collaboration.

Scientific research, in short, is an art; it is the art of making certain kinds of discoveries. The scientific profession as a whole has the function of cultivating that art by transmitting and developing the tradition of its practice. The value which we attribute to science–whether its progress be considered good, bad, or indifferent from a chosen point of view–does not matter here. Whatever that value may be it still remains true that the tradition of science as an art can be handed on only by those practicing the art. There cannot therefore be any question of another authority replacing scientific opinion for the purposes of this function; and any attempt to do so can result only in a clumsy distortion and–if persistently applied–in the more or less complete destruction of the tradition of science.

Professional scientists form a very small minority in the community, perhaps one in ten thousand. The ideas and opinions of so small a group can be of importance only by virtue of the response which they evoke from the general public. This response is indispensable to science, which depends on it for money to pay the cost of research and for recruits to replenish the ranks of the profession. Clearly science can continue to exist on the modern scale only so long as the authority that it claims is accepted by large groups of the public.

Why do people decide to accept science as valid? Can they not see the limitations of scientific demonstrations–in the preselected evidence, the preconceived theories, the always basically deficient documentation? They may see these shortcomings, or at least they may be made to see them. The fact remains that they must make up their minds about their material surroundings in one way or another. Men must form ideas about the material universe and must achieve definite convictions on the subject. No part of the human race has ever been known to exist without a system of such convictions and it is clear that their elimination must mean intellectual death. Without them man falls to the level of the beast as regards both the state of his mind and the level of his technical achievements. That must remain out of the question. The choice therefore open to the public is only that of believing in science or else in some rival explanation of nature, such as that offered by Aristotle, the Bible, astrology, or Christian Science. Of these alternatives the public of our times has in its majority chosen science; and it is the basis of this choice that concerns us here.

Historically, the origin of the decision is not difficult to trace. There were two main battles, one that opened in the sixteenth century against Aristotelian mechanics and astrology, and another that opened in the nineteenth century against the cosmology of the Bible. Both of these led to long-drawn campaigns in which the ideas of science spread rapidly, generation by generation, and finally extended their influence over all the peoples led by the West.

How was this result achieved? It was favored in the first instance by the whole movement of the Renaissance, which aroused independent judgment among educated people. This awakening weakened the forces opposing the dissemination of science. In these circumstances the convincing power of science proved greater than that of its rival. When Galileo demonstrated that objects of widely different weights, when dropped simultaneously from the tower of Pisa, all reached the ground at the same moment, this proved to every witness of the experiment that Aristotle was wrong in teaching that such bodies fall at different rates proportional to their weights. These practical tests were of the same kind as those used by people engaged in the various crafts of mining, building, and in the arts of war; people willing to think for themselves could not fail to be impressed by them. Though scientific proof was not completely accessible to the layman, what he could be shown proved much more convincing to him than the rival arguments based on Aristotle, the church fathers, astrology, or the Bible; and this has continued to hold to this very day.

This does not mean that the victory of science is either complete or final. Pockets of antiscientific views persist in various forms. Scientific medicine is rejected by that part of the public in Western countries which professes Christian Science. Fundamentalism challenges geology and evolution. Astrology has a more or less vague ascendancy in wide circles. Spiritualism carries on a borderline existence between science and mysticism. These persistent centers of heterodoxy are a constant challenge to science. It is not inconceivable that from one of these there may emerge in the future some element of truth inaccessible to the scientific method, which may form the starting point of a new interpretation of nature. In any case, at present these antiscientific movements constitute an effective check on the popular acceptance of science; the failure of their efforts to spread their doctrines shows that science remains considerably more convincing than any of the possible alternatives.

So long as this is the case, science could be discredited with the public only by stopping the channels through which it is disseminated and by suppressing at the same time the desire of people to think for themselves. There may be reasons–which may even conceivably be good ones–for doing these things, but the result could obviously be considered only as a distortion or suppression, not as a redirection of the appreciation of science by the people.

I have shown that the forces contributing to the growth and dissemination of science operate in three stages. The individual scientists take the initiative in choosing their problems and conducting their investigations; the body of scientists controls each of its members by imposing the standards of science; and finally, the people decide in public discussion whether or not to accept science as the true explanation of nature. At each stage a human will operates. But the exercise of will is fully determined on each occasion by the responsibility inherent in the action, and hence any attempt to direct these actions from outside must inevitably distort or destroy their proper meaning.

There are two recent instances on record of attempts to break the autonomy of scientific life and to subordinate it to state direction. The one made by National Socialist Germany was so crude and cynical that its purely destructive nature is easily demonstrated. Take the following utterance credibly attributed to Himmler, in which he reproves German scholars who refused to accept as genuine a forged document concerning German prehistory:

> We don't care a hoot whether this or something else was the real truth about the prehistory of the German tribes. Science proceeds from hypotheses that change every year or two. So there's no earthly reason why the party should not lay down a particular hypothesis as the starting point, even if it runs counter to current scientific opinion. The one and only thing that

> matters to us, and the thing these people are paid for by the State, is to have ideas of history that strengthen our people in their necessary national pride.

Clearly Himmler only pretends here—as a mere form of words—that he wishes to readjust the foundations of science; his actual purpose is to suppress free inquiry in order to consolidate a particular falsehood which he considers useful. The philosophical difficulties in the position of science are mentioned only in order to confuse the issue and to cloak—however thinly—an act of sheer violence.

The attempts of the Soviet government to start a new kind of science are on an altogether different level. They represent a genuine effort to run science for the public good and they provide, therefore, a proper test of the principles involved in such an attempt.

We will illustrate the process and its results by the example of genetics and plant breeding, to which governmental direction was applied with particular energy. The intervention of the state in these fields began about the year 1930 and was definitely established by the All Union Conference on the Planning of Genetics and Selection held in Leningrad in 1932. Up to that time genetics had developed and flourished in Russia as a free science, guided by the standards that were recognized in other countries throughout the scientific world. The conference of 1932 decided that genetics and plant breeding should henceforth be conducted with a view to obtaining immediate practical results and on lines conforming to the official doctrine of dialectical materialism, research being directed by the state.

No sooner had these blows been delivered against the authority of science than the inevitable consequences set in. Any person claiming a discovery in genetics and plant breeding could henceforth appeal directly over the heads of scientists to gullible practitioners or to politically-minded officials. Spurious observations and fallacious theories advanced by dilettantes, cranks, and impostors could now gain currency, unchecked by scientific criticism.

An important case of this kind was that of I. V. Michurin (1855-1935), a plant-breeding farmer, who some years earlier had announced the discovery of new strains of plants produced by grafting. He claimed to have achieved revolutionary improvements in agriculture, and to have obtained a striking confirmation of dialectical materialism. The opinion of science, on the contrary, was–and remains–that Michurin's observations were mere illusions, that they referred to a spurious phenomenon, known by the name of "vegetative hybridization," which had been frequently described before. The illusion can arise from an incomplete statistical analysis of the results obtained and may also be occasionally supported by the fact that viruses are transmitted to the graft and its offspring. The occurrence of true hereditary hybridization by grafting would be incompatible with the very foundations of modern biological science, and its existence had definitely been discredited by the formulation of Mendel's laws and the discoveries of cytogenetics.

The denial of Michurin's claims by scientific opinion now lost its force. His work appealed to the practitioner and it conformed to the philosophy imposed by the state. It thus fulfilled both the criteria which had replaced the standards of science. Hence–inevitably–Michurin's work was now given official recognition. The government, in its enthusiasm over this first fruit of its new policy in science, went even further and erected a monument of unparalleled splendor to Michurin. It renamed the town of Koslov and called it Michurinsk (1932).

The breach thus made in the autonomy of science laid the field of genetics and plant breeding wide open to further invasion by spurious claims. The leader of this invasion became T. D. Lysenko–a successful worker in agricultural technique–who expanded Michurin's claims into a new theory of heredity which he opposed to Mendelism and cytogenetics. His popular influence caused hundreds of people without proper scientific training, such as farmers and young students of agriculture, to attempt grafting experiments with the aim of producing "vegetative hybrids." Lysenko has himself described proudly how by the labors of this mass movement vegetative hybrids "poured out like the fruits from the horn of abundance." Aided by claims of this kind, Lysenko gained high recognition for himself by the government. He was appointed a member of the Academy of the U.S.S.R. and made President of the Academy of Agricultural Science of the U.S.S.R. By 1939 his influence had reached the point where he could induce the Commissariat of Agriculture to prohibit the methods hitherto used in plant-breeding stations and to introduce, compulsorily, new ones that were based on his own doctrine of heredity and that were contrary to accepted scientific opinion. In a publication of the same year he even went so far as to demand the final elimination of his scientific opponents by the total abolition of genetics in Russia: "In my opinion," he wrote, "it is quite time to remove Mendelism entirely from University courses and from the theoretical and practical guidance of seed raising."

However, the government hesitated to take the decisive step and a conference was called to clarify the situation. The editors of the journal *Under the Banner of Marxism* acted as conveners, and the proceedings, together with an extensive editorial commentary, were subsequently published in that journal. The reports of this conference form impressive evidence of the rapid and radical destruction of a branch of science clearly caused by the fact that the conduct of research had been placed under the direction of the state. We may note that the government in this case was a particularly progressive one and that it was aiming at solid, reasonable benefits for its own people. It is all the more significant that in spite of this, the result of its action was only to plunge the science of genetics into a morass of corruption and confusion.

The conference which revealed these conditions to the outside observer was presided over by M. B. Mitin (a person unknown to international science and probably a representative of the journal), who in his opening speech outlined once again what the practical and theoretical principles were to which science must conform when under the direction of the Soviet state. "We have no gulf between theory and practice, we have no Chinese wall between scientific achievements and practical activity. Every genuine discovery, every genuine scientific achievement is with us translated into practice, enters into the life of hundreds of institutions, attracts the attention of the mass of people by its fruitful results. Soviet biologists, geneticists, and selectionists must understand dialectic and historical materialism, and learn to apply the dialectic method to their scientific work. Verbal, formal acceptance of dialectical materialism is not wanted."

Academician N. I. Vavilov, internationally recognized as the most eminent geneticist in Russia (as shown by his recent election as Foreign Member of the Royal Society), put the case for the science of genetics. He surveyed the development of this science from its inception and pointed out that not a single author of repute anywhere outside Russia would either doubt the soundness of cytogenetics or would be prepared to accept the existence of so-called "vegetative hybrids."

Such appeals, however, had now become groundless; with the establishment of state supremacy over science, the authority of international scientific opinion had been rendered void. Vavilov was rightly answered by being confronted with his own declaration made at the Planning Conference of 1932 in which he had depreciated the cultivation of science for its own purposes. Yielding at the time perhaps to pressure, or believing it wise to meet popular tendencies halfway, little expecting in any case the far-reaching consequences to follow from his relinquishment of principles, he had then allowed himself to say: "The divorce of genetics from practical selection, which characterizes the research work of the U.S.A., England, and other countries, must be resolutely removed from genetics-selection research in the U.S.S.R."

Such principles having now been generally accepted, Vavilov could raise no legitimate objection if the classical experiments to which he referred, and on which his branch of science was based, were laughed to scorn by men like the practical plant breeder V. K. Morozov, who addressed the meeting as follows: "The representatives of formal genetics say that they get good 3:1 ratio results with Drosophila. Their work with this object is very profitable to them, because the affair, as one might say, is irresponsible... if the flies die, they are not penalized." In Morozov's opinion, a science which in 20 years had produced no important practical results at his plant-breeding station could not possibly be sound.

This view can in fact be considered as a correct conclusion from the criteria of science now officially accepted (though fortunately by no means universally enforced) in the Soviet Union. If all the evidence drawn from practically unimportant cases is to be disregarded or at least treated lightly, then little proof can remain in support of the theories of genetics. In such circumstances any simple, plausible ideas such as the fallacies advocated by Lysenko must inevitably acquire the greater convincing power and gain the wider support among all nonspecialists, whether practitioners or ordinary laymen. This is in fact what the Conference on Genetics demonstrated. Morozov could assure Lysenko that nearly all practical fieldworkers, agronomists, and collective farmers had become followers of his doctrine of heredity.

The authority of science having been replaced by that of the state, it was also logical that political arguments should be used against Vavilov's traditional scientific reasoning. Lysenko, for example, introduced such arguments as the following:

> N. I. Vavilov knows that one cannot defend Mendelism before Soviet readers by writing down its foundation, by recounting what it consists of. It has become particularly impossible nowadays when millions of people possess such a mighty theoretical weapon as "The Short Course of the History of the All-Union Communist Party (Bolshevists)." When he grasps Bolshevism, the reader will not be able to give his sympathy to metaphysics, and Mendelism definitely is pure, undisguised metaphysics.

It was logical again that Lysenko and his adherents should invoke Michurin as an authority whose claims had been established by the State; that Lysenko should speak of "that genius of biology I. V. Michurin, recognized by the Party and the Government and by the country," and declare that it is "false and conceited" on the part of a biologist to think that he could add anything to Michurin's teaching.

In such circumstances there seems indeed nothing left to the hard-pressed scientists but to attempt a defense in the same terms as used by their opponents. This is what the eminent geneticist Professor N. P. Doubinin apparently decided to do at the Conference on Genetics. His speech in defense of cytogenetics refers freely to Marx, Engels, and the "Short Course of the History of the Communist Party." He reverently mentions Michurin, naming him as a classic next to Darwin. But in his view—as he explains—all these high authorities are directly or indirectly supporting Mendelism. "It is quite wrong," he says, "to describe Mendelism by saying that its appearance represents a product of the imperialist development of capitalist society. Of course after its appearance Mendelism was perverted by bourgeois scientists. We know well the fact that all science is class science."

Such is the last stage in the collapse of science. Attackers and defenders are using the same spurious and often fanciful arguments, to enlist for their own side the support of untutored practitioners and of equally untutored politicians.

But the position of the defenders is hopeless. Science cannot be saved on grounds which contradict its own basic principles. The ambitious and unscrupulous figures who rise to power on the tide of a movement against science do not withdraw when scientists make their last abject surrender. On the contrary, they stay to complete their triumph by directing against their yielding opponents the charge of insincerity. Thus Lysenko says, "The Mendelian geneticists keep silent about their own radical disagreement with the theory of development, with the teaching of Michurin," and even more jeeringly is the same taunt made by Lysenko's assistant Professor I. I. Prezent: "It is new to find that all of them, some more sincerely than others, all of them try to give the impression that with Michurin at least they have no quarrel."

Such taunts are unanswerable and their implications are shattering. They make it clear that scientists must never hope to save their scientific pursuits by creeping under the cloak of essentially antiscientific principles. "Verbal, formal acceptance of these principles," the Chairman had sternly warned from the beginning, "is not wanted."

The demonstration given here of the corruption of a branch of science caused by placing its pursuit under the direction of the state is, I think, complete. The more so, I wish to repeat, as there is no doubt at all about the unwavering desire of the Soviet government to advance the progress of science. It has spent large sums on laboratories, on equipment, and on personnel. Yet these subsidies, we have seen, benefited science only so long as they flowed into channels controlled by independent scientific opinion, whereas as soon as their allocation was accompanied by attempts at establishing governmental direction they exercised a violently destructive influence.

We may hope and expect that one day the Soviet government will recognize the error in such attempts: that they will realize, for example, that their plant-breeding stations are operating on lines which were abandoned as fallacious in the rest of the world about 40 years earlier.

What can a government do when it realizes such a state of affairs? What course can it then take to restore the functions of science?

According to our analysis the answer cannot be in doubt. One thing only is necessary—but that is truly indispensable. It is only necessary to restore the independence of scientific opinion. To restore fully its powers to maintain scientific standards in respect of all their proper functions, in the selection of papers for publication, in the selection of candidates for scientific posts, in the granting of scientific distinctions, and in the award of special research subsidies. To restore to scientific opinion the power to control by its influence the publication of textbooks and popularizations of science as well as the teaching of science in universities and schools. To restore to it above all the power to protect that most precious foothold of originality, that landing ground of all new ideas, the position of the independent scientist—who must again become sole master of his own research work.

There is still time to revive the great scientific tradition of Russia which, although at present distorted in many respects, is very far from being dead. The recent great progress of Russian mathematics, and of many other fields in which state control has never been effectively applied, proves that the valuation of science for its own sake still lives in U.S.S.R. Let scientists be free once more to expound their true ideals. Let them be allowed to appeal to the Soviet peoples; to ask for their support of science on its own grounds, for the understanding and love of science when pursuing its own immortal purpose. Let them be free to expose the cranks and careerists who have infiltrated their ranks since the inception of "planning" in 1932. Let the Soviet scientists become affiliated again to the body of international science.

The very moment that scientists regain these freedoms, science will flourish again. Overnight it will rise again free of all the confusion and corruption which is now affecting it; and, aided by the rich endowments which it is receiving from the state, the Russian scientific genius will once more speed on, unhampered, to new great achievements.

However, the current of future events may well tend toward the very opposite course. We are experiencing today a weakening of the principles of scientific autonomy in countries where science is still free. There is a movement afoot among scientists themselves, urging that science should be adjusted to social ends. "Science must be marshaled for the people" Professor H. Levy is reported to have proclaimed at a recent popular rally of scientists in London. Fired by misguided generosity, these scientists would sacrifice science–forgetting that it is theirs only in trust for the purpose of cultivation, not theirs to give away and allow to perish.

Our analysis seems to leave no doubt that if this kind of movement prevailed and developed further–if attempts to suppress the autonomy of science such as have been made in Russia since 1932 became world-wide and were persisted in for a time–the result could only be a total destruction of science and of scientific life. Single individuals could perhaps continue to study pure science and to achieve sporadic progress as some did even during the Middle Ages. But the swift and steady progress of discovery, as experienced in the past 100 years, would be brought to a standstill and soon the main body of science itself would disintegrate and fall into oblivion.

That is why we must recognize the essentially autonomous nature of science with all its great implications.

2

SCIENCE AND THE MODERN CRISIS

History will view the events which have taken place on the Continent during the last generation as one coherent process of upheaval. The rise of a totalitarian regime in Russia and the growth of Fascism in other European countries will be seen to arise from joint sources. These movements will then represent together the breakdown of a previous system of public life and its replacement by disastrous innovations.

The form of European civilization which was submerged by the European upheaval still stands out clearly in the memory of those who, like myself, came to maturity before the last war. It was a liberal civilization in which free institutions had been established throughout Europe and were still being actively developed in most places. Even in the most oppressed parts of the Continent, like Czarist Russia, liberal development had been continuously going on throughout the nineteenth century and had already achieved considerable success. To take only one feature, recall the activities of the Russian Duma, a kind of parliament which existed during the last 11 years before the outbreak of the Revolution. Through most of this time the Duma was dominated by parties which were sharply opposed to the existing autocractic form of government. The Marxist Social Democratic party, for example, which had been formed 30 years earlier, was represented by a considerable number of members. Though constitutionally pow-

First published in *Memoirs and Proceedings of the Manchester Literary and Philosophical Society*, Sessions 1943-45, 86 (June):7-16, 1945.

erless, the Duma did actually exercise political influence, and even succeeded in overthrowing governments to which it objected.

The degree of freedom and tolerance which were generally taken for granted in Europe before the European upheaval is inconceivable to the present generation. Suffice it to say that the very conception of totalitarianism was unknown—except as a matter of speculation—before 1917 in Europe. No secular authority had even remotely attempted to enforce among its citizens a conformity of views such as is now commonly demanded in totalitarian countries.

Apart from its intolerance, modern totalitarianism exhibits as its characteristic a new, hardheaded materialist conception of politics. Both Marxism and Fascism conceive of public affairs in terms of force and force alone, and express sweeping contempt for the ideals which nineteenth-century politicians pursued or professed to pursue. And as a further characteristic we find—rather paradoxically combined with this hardheaded realism—a new fanaticism, unknown to Europe in the liberal era. Thus intolerance, realism, and fanaticism now produce together that harsh and somber uniformity which has come to replace in most of Europe the easy and varied affairs of the nineteenth century.

Naturally, we must never forget the sharp differences between the two strains of the European upheaval; the Marxist Revolution which pursues, or has pursued for most of the time, universalistic aims intended to benefit all mankind, and the Fascist movements which set themselves the limited purpose of aggrandizing one nation at all costs. But in spite of this distinction we may regard the joint European upheaval, and the further spread of its common ideas, as constituting the crisis of our times.

Some writers would explain this crisis by a partial breakdown of capitalism. But this supposition is clearly unsound, as it fails to explain why the Marxist Revolution broke out in a country in which industrial capitalism had hardly developed at all. Other interpretations of the European crisis are based on spiritual or, generally, mental grounds, and my own comments will be along this line, as my subject is to examine the part which the rise of modern scientific thought played in producing the disasters.

Historically—and I believe also logically—the origins of the modern crisis can be traced back to the very beginnings of modern civilization as it emerged from the Middle Ages. When the power of feudal hierarchy was gradually undermined and increasingly displaced by the authority of the state, this involved an emancipation of public authority from the tutelage of the Church and the establishment instead of a new supreme authority based on secular foundations. The political philosophy of the time eagerly realized the problems involved in this change. Speculations concerning the nature of the new state and the proper justification and measure of its powers were pursued by many thinkers. A weighty contribution to this discussion was made by the English philosopher Thomas Hobbes in his book *Leviathan,* published in 1651, which did in fact provide an early formation of totalitarian doctrines.

Hobbes was profoundly affected by the teachings of Galileo, according to which the visible universe consists of matter in motion, and he thus became the father of scientific materialism. As a pioneer of materialism, he was particularly suited to develop to its utmost logical implications the conception of a state based on purely secular foundations. He started to develop this conception from the assumption that the state itself was subject to no obligations of any kind; that it was in fact absolute master of its own destiny. Next he demanded that the state, if properly constituted, should be in a position to avert civil war between its citizens, that it should have power to keep the peace in all emergencies within its realm. From this he derived further the momentous conclusion–famous in the history of political thought–that all ultimate power must be vested at one single point of the state. Any division of ultimate power, he argued, would leave open the possibility of a conflict between its parts, and since, by assumption, there would exist no authority superior to both parties, such conflict would represent a state of actual, or at least latent, civil war. Hence, he said, no division of powers is permissible. The logic of this argument seems irrefutable. If the state radically refuses to acknowledge any higher power over itself, then it must be invested with absolute powers over all its subjects. A sovereign must then be free of any obligation to respect the law and be entitled to override equally any objections raised in the name of religion and morality, and even any claims of plain truth. The principle must be accepted that the sovereign can do no wrong.

Like his modern totalitarian successors, Hobbes reassures his audience that the absolute powers of the state would be all to the good. An absolute ruler, he maintained, is so identified with his subjects that he cannot but desire their welfare; and in any case there can be no welfare for them outside their dependence on the sovereign. By the same reasoning Hobbes disposed of any claims of religion, morality, and science which might get into conflict with the sovereign's decisions. He argued that no statement could be true if its contents ran contrary to the welfare of the state, as represented by the sovereign. Thus he anticipated the most modern form of totalitarian theory, according to which the test of truth lies in its usefulness to society, the interests of which are properly represented by the authority of the state.

However, all these extraordinary teachings of Hobbes pretty much remained on paper, at least so far as English political life in Hobbes's time was concerned. Hobbes's logic was impeccable, but his assumptions were not fulfilled in reality. There *were* superior powers in existence at the time, and very effective powers at that, to which any sovereign ruler of England had to defer. The Bible was one such power. The seventeenth century was filled with religious fervor which had to be respected by the state. In these circumstances the position assigned by Hobbes to religion under the state was mere bookish speculation, unrelated to the realities governing men's minds. So long as religion and the great human ideals of justice, morality, and truth–which at the time were so closely related to religion–continued to be respected in their own right, the totalitarian logic of Hobbes could take no effect.

The power exercised in the public affairs of England by religious beliefs and by human ideals was actually growing stronger in Hobbes's time. A great Protestant and humanitarian movement was in the ascendant and was taking over the guidance of political progress in England for times to come. This movement was in fact to develop those institutions of tolerance and self-government which eventually were to spread from England to America, as well as to France and other parts of the Continent, and which formed, in the nineteenth century, the foundations of the great liberal civilization of that time.

However, in spite of the progress of these events, the logic of Hobbes in due course got a chance to assert itself in France, through the decline of religious beliefs in that country. The rise of modern science had given birth to French rationalism. Under Voltaire's leadership the French Enlightenment had rebelled against religious dominion. Agnosticism or deism was becoming generally accepted among educated and progressive people in Europe. The collapse of religious beliefs had led to the release of new humanitarian aspirations of a secular kind. These new aspirations were arousing demands for political reform. They were swelling the tide of resentment against the ruling semifeudal system and fermenting the political unrest which was to culminate in the French Revolution.

Great hopes which had hitherto found expression in religion were thus for the first time attached to the outcome of politics. The political philosophy which presided over this transformation was that of Rousseau, in whose mind the logic of Hobbes had become active in a novel fashion. Rousseau agreed with every word that Hobbes had said about the sovereign–he reaffirmed that the sovereign must be absolute and can do no wrong–but to this he added a formidable reservation. He declared that a ruler was truly sovereign only if his power was held justly and not by usurpation. The people were the true sovereign; and power was just only if it emanated from the people. Rulers who did not represent the people were usurpers, betrayers of the nations whom they were keeping enchained against their will. Thus the doctrine of Jacobinism was foreshadowed; the doctrine–as Lord Acton worded it–"That a government truly representing the people could do no wrong." We know how this doctrine led to the first attempt at social salvation by tyranny in Europe, to the first regime of political terror, and to the first great political purges–how the Jacobin patriot, believing himself to be the true representative of the nation, felt justified in exterminating his political opponents and rivals as traitors to the people.

However, Jacobinism was a premature attempt at the application of Hobbesian doctrine in European politics. The decay of religious beliefs was not in itself enough to open the way for the operation of the totalitarian logic. Totalitarianism cannot be established safely so long as the ideals of justice, humanity, and truth continue to be respected. In vain did the Jacobins quote Rousseau in support of their actions. Massacres, such as they had perpetrated, were not yet excused as necessary precautions against a potential fifth column. Their aspect roused a wave of revulsion through the world and estranged the warmest friends of the Revolution. In France itself the Jacobin terror collapsed under the burden of its crimes. The moral standards of humanity were still on guard against the advance of the Hobbesian Leviathan, even in its new popular revolutionary guise.

In the course of the nineteenth century the humanitarian ideals which had found an outlet in the French Revolution went on spreading peacefully and gradually transformed the whole of Western civilization. Thus was the great liberal era inaugurated which went on flourishing until the outbreak of the present European disasters. But the axe had been laid to the tree of liberalism from the very beginning of the nineteenth century. On the Continent of Europe scientific materialism was now vigorously advancing once more. But, having destroyed religious beliefs among the leaders of progress, materialism was now beginning to undermine its own belief in the reality of human ideals. A new, entirely naturalistic, conception of man and of human society was becoming generally accepted by the progressive intelligentsia on the Continent.

The movement was strongest in the central and eastern parts of the Continent, particularly in Germany and Russia. It was in Russia that the figure of the modern nihilist made its first appearance. Turgenev described him in the person of Bazarov, hero of his novel *Fathers and Sons,* which was published in 1862. Bazarov denies the reality of all human ideals—even of love—and professes that man is nothing but a bundle of appetites. Turgenev was attacked by the Russian radicals at the time, but he upheld his hero as a true portrait of the Russian intellectual and declared that he himself had arrived at the views of Bazarov.

The new materialist doctrine had come to Russia mainly from Germany, where it had spread by the writings of Büchner, Vogt, and Moleschott. Büchner's famous *Kraft und Stoff,* published in 1855, is represented as the bible of the young generation in *Fathers and Sons.* A characteristic phrase of Büchner's was: *"Ohne Phosphor kein Gedanke"* ("Without phosphorus no human thought"). And no less men than Feuerbach and Karl Marx echoed this outlook by the pronouncement: *"Der Mensch ist was er isst"* ("Man is what he eats").

Marx was imbued with this form of materialism, much more even than his doctrine directly expresses it. His biographer, I. Berlin (1939), tells how the manuscripts of the numerous manifestoes and programs of action to which Marx appended his name still bore the fierce marginal comments with which he obliterated all reference to eternal justice, to the equality of man, the rights of individuals or nations, to the liberty of conscience, to the fight for civilization, and other democratic movements of his time–as he looked on all these as mere cant, spreading confusion and pointless action.

Yet Marx did not become the founder of modern totalitarianism by delivering Hobbes's Leviathan from any moral opposition, as had caused the fall of Jacobinism.

To understand the position we must realise that Marx–in spite of his denial of morality as an independent force–was passionately dominated by moral motives. Even while he was angrily crossing out in his manifestoes all references to social justice and human sympathy, he was burning with those very sentiments and rousing their fire everywhere among his followers. Moral sentiments were not killed by Marx when he deprived them of independent standing–but driven underground. They were henceforth to operate in the dark; as the force behind the science of the Marxist movement.

Here lies the origin of modern fanaticism. It is the force of moral aspirations driven underground, and giving their blind support to a supposed scientific theory which promises social salvation by the mere force of violence. When our intellect convinces us, backed by the authority of science, that our morality is pointless, and teaches us that we can achieve everything to which morality aspires merely by letting loose our animal forces–then our morality is converted into scientific bestiality. That is the picture of the modern fanatic, of the modern party man; aloof, and supremely confident of possessing a superior knowledge of reality; cruel and unscrupulous; merciless torture and death.

Thus Marxism not merely makes totalitarian logic secure, but also endows Leviathan with revolutionary powers more fierce even than those which it possessed in its Jacobin form. Marxist materialism enables the supreme secular power postulated by Hobbes to be established in its perfect form. All religion, morality, in fact all forms of human consciousness having lost their independent standing, there is no higher power left to which force would have to defer.

Leviathan attains logical perfection. But in addition, Marxism endows Leviathan with the assurance of science and the appetites of the masses, and inflames the monster with a fanaticism distilled from the highest aspirations of mankind.

I can now deal briefly with the Fascist movements of our time. Fascist doctrines are constructed on the same plan as Marxism but they differ from the latter in their moral content. The primary motive incorporated in Fascism is a sentiment of patriotism. This is a much narrower aspiration than the desire for universal justice which underlies Marxism, but it is yet of a high moral order; true patriotism includes a sense of national obligation and of national honor.

But Fascism divests patriotism of its humane and honorable attributes. Fascist doctrine, and the Nazi doctrine in particular—which is the most highly developed form of Fascism—transforms patriotism into a purely materialist force. It converts patriotism from an ideal into a theory of violence. We have here the counterpart of the transformation of socialism by Marx "from a Utopia to a science."

It is a mistake to regard the Nazi as an untaught savage. His bestiality is carefully groomed by speculations closely reflecting Marxian influence. He holds a theory of national salvation by violence, based on the preponderance of the German race on the Continent, which is at least as well founded in reality as the theory of class war. His contempt for humanitarian ideals has a century of materialist schooling behind it. It goes back to the same origin as Marx's hatred of moral arguments—and, for that matter, Nietzsche's similar hatred of morality. The Nazi disbelieves in public morality in the way we disbelieve in witchcraft. It is not that he has never heard of it, but that he thinks there are positive grounds for asserting that such a thing cannot exist. If you tell him the contrary, he will think you peculiarly old-fashioned, or simply dishonest. Nor is Nazi totalitarianism based on any peculiar devotion of the German people to the state. It rather represents the plain logic of all purely temporal power. It is simply the Leviathan of Hobbes enthroned and made unassailable by modern scientific materialism.

Turning back, in conclusion, from the scene of the European upheaval to events as they have developed in England since the days of Hobbes, we see now even more clearly that they belong to an altogether separate branch of European civilization. In Britain the logic of the Leviathan has remained in suspension—not because of any magic which can prevent logic from fulfilling its implications in British public life (which is a pernicious illusion), but simply because scientific materialism was not accepted by the leaders of progress in Britain. Religion retained a dominant position in the public life of the English-speaking countries and moral arguments retained their position in the guidance of public policy. Had this been otherwise, or were it ever to become otherwise, the logic of Leviathan would come to be fulfilled in this country exactly as it has been fulfilled elsewhere.

I have not tried to explain why the effect which science had on the outlook of particular people was more radical than its effect on others: why Newton's discoveries kindled a new nationalist Enlightenment in France and not in Newton's own country. Why again in the nineteenth century a new materialism was derived from science on the Continent of Europe and not in the English-speaking countries; and why this materialism was implicitly accepted in the progressive circles of Germany and Russia and not to the same extent, say, in France and Holland. Having offered no views on these questions I cannot propose to discuss now the prospects of scientific materialism in Britain. But it is obvious that a new, hardheaded, utilitarian outlook spread vigorously in the interwar period both here and in America. If the conception of the crisis of our time which I have put forward here is even approximately correct, the fateful potentialities of such a change should be clear.

3

THE SOCIAL MESSAGE OF PURE SCIENCE

I have been asked to talk about fundamental science in its relation to the community. This relation was rarely discussed until about 10 years ago, but since then, rather suddenly, it has become a major problem. What justification is there in scientific studies having no visible practical use? Until fairly recently it used to be assumed that such studies served their own purpose, namely the advancement of knowledge–the discovery of knowledge simply for the love of truth. Do we still fully accept that view? Do we still believe that it is proper for a scientist to spend public funds for the pursuit of studies such as, say, the proof of Fermat's theorem, or the counting of the number of electrons in the universe? Studies which, though perhaps not lacking in some very remote possibility of practical usefulness, are at any rate as unlikely to yield a material dividend as any human activity, within the realms of sanity, could possibly be? No, we do not generally accept today, as we did until the 1930s, that it is proper for science to pursue knowledge for its own sake, quite regardless of any advantage to the welfare of society. Never was there more talk about science—never less heard about the love of knowledge. We are faced here with a turn to a radically utilitarian attitude reflecting the general philosophical trend of our time.

The philosophical movement which has so effectively impaired the traditional standing of science has been attacking it from two different sides. One line of attack was directed against the claim of

First published in *Nineteenth Century*, 146 (July):14-27, 1949.

science to speak in its own right. This is the line of modern materialistic analysis which denies that the human intellect can operate independently on its own grounds and holds that the purpose of thought is at bottom always practical. Science in this view is merely an ideology, the contents of which are determined by social needs, and the progress of science is seen as depending on a series of successive dominant practical interests. Newton, for example, is then represented as discovering gravitation in response to rising navigational interests, and Maxwell as discovering the electromagnetic field in order to make possible trans-Atlantic communications. Such a philosophy wipes out all distinction between pure and applied science, and it values pure science only for the fact that it may turn out to be useful in the end.

The other line along which the new philosophy attacks the position of science is based on moral grounds. It insists that scientists should turn their eyes to the misery which fills the world and think of the relief they could bring to it. It asks whether on looking round they can find it in their hearts to use their gifts for the elucidation of some abstruse problem—as for example the counting of the electrons in the universe, or the solution of Fermat's theorem. Could they possibly prove to be so selfish? In this manner scientists are reproached for pursuing science for the mere love of knowledge.

Thus we can see the position of pure science today under the crossfire of two attacks based on rather disparate grounds, forming a somewhat paradoxical combination—but one that is actually very typical of the modern mind. A new destructive skepticism is seen united to a new passionate social conscience; an utter disbelief in the spirit of man is seen coupled with extravagant moral demands. We see at work here the mode of action which has already dealt so many shattering blows to the modern world: the chisel of nihilism driven by the hammer of social conscience.

This is the lesson to be read from the scene of Europe. The destruction of our civilization over large stretches of the Continent was not due to an accidental outbreak of Fascist beastliness. The events which, starting from the Russian Revolution, have ravaged the Continent represent on the contrary one vast coherent process of upheaval. Great waves of humanitarian and patriotic sentiments were its prime impulses, and it was these sentiments which released destruction over Europe. Savagery is always there lurking among us, but it can break loose on a grand scale only when moral passions first smash up the controls of civilization. There are always some potential Hitlers and Mussolinis about, but they can gain power only if they succeed in perverting moral forces to their own ends.

We must ask, therefore, why moral forces could be thus perverted; why the great social passions of our time turned so readily into violent and destructive channels. The reason seems clearly that there was no other channel available to them. A radical skepticism had destroyed popular belief in the reality of justice and reason. It had stamped these ideals as mere superstructures, as out-of-date ideologies of a bourgeois age, as mere screens for the interests hiding behind them, and indeed, as pitfalls of confusion and weakness to anyone who would trust in them. No longer was there a sufficiently strong belief in justice and reason in which to embody social passions. And there grew up on the Continent a generation full of moral fire and yet despising reason and justice. Believing instead—in what? Well, in what there was left for them to believe in. In power, economic interest, subconscious desire. Such were the forces left for them to rely on, and these they accepted as the ultimate reality to which they could entrust themselves. Here they found a modern, hard-boiled, tough embodiment for their moral passions. And the highest hopes of humanity were thus embodied in violence. Social sentiment was turned into hatred and the desire for the brotherhood of man into deadly class war. Patriotism was turned into Fascist beastliness; the more beastly the more patriotic the people who had gone Fascist.

I may quote here Mr. Atlee describing the most urgent need in Europe at the present time. "We need," he said, "a conception of justice not as a will of a section, but as something absolute," and a leadership "which will lift people up from a mere longing for material benefits to a sense of the highest mission of mankind." Mr. Bevin has spoken in a similar fashion when, facing the starving masses of Europe, he talked of a "spiritual hunger which is even more devastating than physical hunger."

But alas, the doctrine which had been so effectively hammered into our heads by the leading philosophical movement during the last 15 years had taught us precisely this: that justice is nothing but the will of one section, and that there can be nothing higher than the longing for material benefits; so that to talk about higher missions is just foolishness or deceit. No, the spiritual hunger of Europe will not be satisfied so long as we follow the leadership of those–whether on the left or on the right–who teach that material interests alone are real. The most urgent need of the day is to oppose this philosophy with all our might and at every point. To us scientists it falls to attack it in connection with science. The most vital service we owe to the world today is to restore our own scientific ideals which have fallen into discredit under the influence of the modern philosophical movement. We must reassert that the essence of science is the love of knowledge and that the utility of knowledge does not primarily concern us. We must stand up and demand once more for science that public respect and support which is due to it as a pursuit of knowledge and of knowledge alone. We must reassert the claims of science as part of the claims of scholarship and demand respect for all scholarship. We must also re-establish the claims of academic independence and of academic freedom. We scientists are pledged to a higher obligation, to values more precious than material welfare; to a service far more urgent than that of material welfare. Europe can be saved only by the spirit. Our duty is to keep faith with the spirit in science.

How sharply the spirit of pure scholarship is opposed to the claims of totalitarianism has been sufficiently proved on many cruel and terrible occasions in contemporary history. Universities which upheld the purity of their standards under totalitarianism invariably had to stand up to harsh pressure and often to suffer heavy penalties. The whole world recognizes today its debt to universities in Poland and Norway, in Holland, in Belgium and France, where such pressure was withstood and such penalties were endured. These places are witnesses today to the convictions underlying our European civilization and hold out the hope of a genuine European recovery. And where, on the contrary, universities have allowed themselves to be bamboozled or terrorized into compromising their standards, we feel that the very roots of our civilization have been marred. In such places our hopes for the future burn dark and low.

The world needs science today above all as an example of the good life. Spread out over the planet scientists form even today, though submerged by disaster, the body of a great and good society. Even today the scientists of Moscow and Cambridge, of Bangalore and San Francisco, respect the same standards in science; and in the depths of shattered Germany and Japan a scientist is still one of ourselves, upholding the same code of scientific work. Isolated though we are today from each other, we still bear the mark of a common intellectual heritage, and claim succession to the same great forerunners.

We must demand the revival of international scientific life, as part of the restoration of reason and civilized human intercourse in Europe. We need no official delegates nor the help of any official agencies. Let the United Nations simply extend the privileges of diplomatic passage–or at least the privileges of the press–to scientists and scholars all over the world. Let them but give us scientists freedom to travel as we please through Europe and we shall restore within six months a close and intensive collaboration of all scientific workers in Europe. We shall reweave once more at least this part of the destroyed European fabric. By a single cut through red tape a vital aspect of European civilization would be restored to life once more.

This we must demand; and if governments will not listen to us, we shall at least have made our position clear and formed a rallying center for other forces advancing on parallel lines. We shall stand ready to support other demands for the resumption of European intercourse and the restoration of European standards. If such demands sound utopian today, this only proves how urgent they are; how widely we have fallen away from what even a short time ago were normal requirements of civilized life. It only demonstrates once more what dynamic antagonism the claims of our academic tradition can exert against oppressive governments in our times.

Such is my conception of the relation of science to the community in our times. In the great struggle for our civilization science occupies a section in the front line. The movement which is undermining the position of pure science is, in my view, one detachment of the forces assailing our civilization. I know perfectly well that these forces embody some of the most penetrating thoughts and some of the most generous sentiments of our days, but that makes it only the more dangerous in my eyes. We shall have to fight, in this battle, to our sorrow, some of the best motives of human progress. But we cannot afford to turn aside for them. The easy wisdom of the modern skeptic, robbing him and his followers of spiritual guidance and setting free so much untutored enthusiasm, has cost us already too dearly. Whatever scorn be poured upon us by some who find our faith in pure science old-fashioned, and whatever condemnation by others who will think us selfish, we must persist in vindicating the ideals of science. We must fight for their complete restoration to their original position.

Europe is in mortal danger today through a state of all-pervasive suspicion. On the Continent few political parties feel safe with their opponents holding the ministries of police or war. In countries under dictatorship any opposition is rigidly presumed to be treason. The aggregate of all these suspicions is embodied on a colossal scale in the tension between Russia and the West. Yet I can see little wickedness abroad and much goodwill everywhere. If in spite of this we are in mortal danger from general suspicion, it is because so many people earnestly believe today that only violence can achieve results that are worth while. The nobler their purpose the more profound the perversion, and the greater its menace to civilization. The only hope for Europe is that its leaders may become converted once more to a belief in solutions by reason and justice. No triumphs of applied science can help us; no improvement of the atomic bomb, nor even a defense against the bomb; not jet planes outpacing sound; nor new vitamins or penicillins–not even with the remedies for cancer and tuberculosis added to them. But wherever we reassert the spirit of pure science and pure scholarship we shall also help to revive the spirit of reason and justice, and shall thus supply an element which is indispensable to European survival and make a start along the only path toward recovery and safety. We may not succeed. But even so, at least, by proclaiming our true philosophy we shall face the disasters of our time in a manner which may transmit to later ages the true mission of science.

4

THE NATURE OF SCIENTIFIC CONVICTIONS

I

There are many old jokes about the futility of philosophizing. It is quite true that science is a much more businesslike occupation, in which every achievement, however modest, may give you sound satisfaction. Your work stands there, public, compelling, and permanent; it testifies that for one moment you were allowed to make intellectual history. You have disclosed something that had never been known before and that–you may hope–will henceforth go on to be known as long as the memory of our civilization endures.

Some philosophers of the last century were so much impressed by this kind of positive achievement that they decided to liquidate philosophy altogether and divide up its subject matter among different sciences. A number of new sciences, which took man or human affairs as their object, appeared to become available at the time for this purpose. Psychology and sociology were acclaimed as the principle legatees in this sharing out of the substance of philosophy.

This philosophy-to-end-all-philosophy may be designated, if somewhat loosely, as positivism. It continued in the nineteenth and twentieth centuries the rebellion against the authority of the Christian churches, first started in the days of Montaigne, Bacon, and Descartes. But the movement set out not only to liberate reason from enslavement by authority, but also to dispose of all traditionally guiding ideas, so far as they are not demonstrable by science. Thus, in the positivist sense truth becomes identified with

First published in *Nineteenth Century,* 146 (July): 14-27, 1949.

scientific truth and the latter tends–by a positivist critique of science–to be defined as a mere ordering of experience. In this light, justice, morality, custom, and law appear as mere sets of conventions, charged with emotional approval, which are the proper study of sociology. Conscience is identified with the fear of breaking socially approved conventions and its investigation is assigned to psychology. Aesthetic values are related to an equilibrium of opposed impulses in the nervous system of the beholder.[1]

In the positivist theory, man is a system responding regularly to a certain range of stimuli. The prisoner tortured by his jailers in order to extract from him the names of his confederates, and similarly, the jailers torturing him for this purpose, are both merely registering adequate responses to their situations.

Under the guidance of such concepts, we are expected to become truly detached and objective in our approach to the whole wide world, including our own selves and all the affairs of men. Scientific man will thus master both his inner conflicts and those of his social environment and, set free from metaphysical delusions, will henceforth refuse to submit to any obligations that cannot be demonstrated to lie in his proper interest.

Such a program implies of course that science itself is "positive," in the sense that it involves no affirmation of personal beliefs. Since this is in fact untrue–as I shall try to show at some length here–it is not surprising that the positivist movement, having first exalted science to the seat of universal arbitrament, now threatens to overthrow and destroy it. I regard the tension between Marxism and science, which has made its appearance in Soviet Russia and has become steadily more intense during the past 15 years, as a manifestation of this threat, and see in it a logical consequence of the conflict between the aspirations of positivism and the true nature of science.

II

We shall get our attitude to science into better perspective if we digress for a moment on some forms of knowledge which are rejected by science and which most of us hold to be erroneous. Take

[1]Only the last item on this list requires supporting evidence, for which see I. A. Richards (1924), e.g., pp. 245, 251.

sorcery and astrology. I shall assume that both are held to be false by the reader; but the same obviously does not hold for everybody even today. Sorcery, for example, is being practiced among primitive people throughout the planet. In order to bewitch somebody, the sorcerer gets hold of an appurtenance of his victim–such as a lock of hair, a bone he had spat out, or any excretion of his–and burns this object, pronouncing a curse on its owner. This is believed to be effective, and it is common among primitive communities even to ascribe all manner of death to such sorcery.

Now if we ask "What is sorcery?" we cannot say that "It is the destruction of human beings by burning a lock of their hair," etc. For we do not believe that men can be killed by such means. We have to say "There is a belief in sorcery–which we do not share–and which affirms the possibility of killing a man by burning, for example, a lock of his hair." And we cannot define astrology as a method for predicting the course of men's lives by casting their horoscopes, but can only describe it as a belief–which we do not share–in the possibility of foretelling the future from the stars.

Naturally, a sorcerer or an astrologer would speak differently. The first may state that sorcery is the way of killing a man by burning a lock of his hair, or the like; the second will describe astrology as the art of predicting the future from horoscopes. However, if pressed by our skepticism, they would no doubt be prepared to recast their accounts of sorcery or astrology into a statement similar in form to our own definition, while replacing the words "a belief which we do not share" by "a belief which we share." And on these grounds we could both agree to differ.

All this has its application to science. Any account of science which does not explicitly describe it as something we believe in is essentially incomplete and a false pretense. It amounts to a claim that science is essentially different from and superior to all human beliefs that are not scientific statements–and this is untrue.

To show the falsity of this pretension, it should suffice to recall that originality is the mainspring of scientific discovery. Originality in science is the gift of a lonely belief in a line of experiments or of speculations which at the time no one else considered to be profitable. Good scientists spend all their time betting their lives, bit by bit, on one personal belief after another. The moment discovery is claimed, the lonely belief, now made public, and the evidence produced in its favor, evoke a response among scientists which is another belief, a public belief, that can range over all grades of acceptance or rejection. Whether any particular discovery is recognized and developed further, or is discouraged and perhaps even smothered at birth, will depend on the kind of belief or disbelief which it evokes among scientific opinion. Let me show how this works, or has worked in some instances. Take the reception accorded to two papers published by two authoritative scientists in Britain at about the same time. One of these was published in the *Proceedings of the Royal Society* in June, 1947, by Lord Rayleigh, a distinguished Fellow of the Society. It described some simple experiments which proved, in the author's opinion, that hydrogen atoms impinging on a metal wire could transmit to it energies ranging up to 100 electronvolts. Such an observation, if correct, would be of immense importance—far more revolutionary, for example, than the discovery of atomic fission by Otto Hahn in 1939. Yet when this paper came out and I asked various physicists' opinion about it, they only shrugged their shoulders. They could not explain the results stated; yet not one believed in it, or even thought it worth while to repeat the experiment. They just ignored it. Since Lord Rayleigh has since died, the matter seems to have been already forgotten.

About simultaneously with Lord Rayleigh's paper Professor P. M. S. Blackett published (May, 1947) the fact that a simple relationship between angular momentum and stellar magnetism was applicable to the earth, the sun, and a third star, the data for which lie over a wide range of values. This communication, though meager as compared with Rayleigh's and not of obvious significance, was received as an important discovery which justified every effort toward its further exploration.

Now I feel sure that 30 years earlier the reaction would have been exactly the reverse. Before the discovery of general relativity, the kind of relationship suggested by Blackett would have been shrugged aside as just one more curious numerical coincidence, of which there were so many; while Lord Rayleigh's observations would have been acclaimed at their face value, since they would not have been clearly incompatible with the theories current at the time on the nature of atomic processes.

We can see here the vital function exercised by current beliefs as to the nature of things on the course of scientific development. It may well turn out that scientific opinion has misplaced its beliefs in one or even in both of the instances I have given. Yet this would be no reason for refusing to consider such fiduciary decisions, since without them science could not operate at all. This has always to be borne in mind when meeting serious errors made by scientific opinion in the face of some alleged new discoveries.

A memorable example of such an error is offered by the history of hypnotism. The process today called hypnosis seems to have been known among nonscientific people from earliest times. The potency of curses among primitive tribes may be due to it; the practices of Hindu fakirs are examples of hypnosis; and many magical performances, and some reputed Christian miracles as well, can now be explained in terms of hypnosis. However, our fundamental beliefs of science had first arisen in direct opposition to beliefs in sorcery and miracles, and the ancient facts of hypnotism found, therefore, no place in the new scientific outlook. They were ignored along with the innumerable superstitions which science had come to supersede. When the facts were once more brought to light by various scientists about two centuries ago, their observations were quietly ignored by science. Then, toward the end of the eighteenth century, the matter was brought to a head by the public demonstrations of Friedrich Anton Mesmer, a Viennese medical practitioner, whose hypnotic cures had spread his fame all over Europe. Scientific commissions repeatedly investigated the facts produced by Mesmer and either denied them or explained them away. Finally, Mesmer was broken, his art discredited, and he himself stigmatized as an impostor. A generation later another pioneer of hypnotism, Elliotson, a professor of medicine at the University of London, was ordered by the university authorities to discontinue his hypnotic experiments; whereupon he resigned his chair. At about the same time, Esdaile, a surgeon in the service of the government of India, performed as many as 300 major operations under hypnotic anesthesia, but medical journals refused to publish his account of these cases. His patients, who had uncomplainingly suffered their limbs to be cut off, were charged with collusion. In one instance (in England) the opponents of hypnotism even produced a "confession" from a patient whose thigh had been amputated—but he later withdrew it. Braid, a medical practitioner of Manchester who took up the matter shortly before Esdaile, was listened to with somewhat lessened hostility, for he started off by attacking the followers of Mesmer and attempting to explain away the process of suggestion. But even Braid's work (which finally did establish suggestion once more) was neglected and ignored for another 20 years after his death. It was not until Charcot took up hypnosis at the Salpetrière in Paris, almost a century after Mesmer's acclamation by the lay public, that hypnotism gained full acceptance among scientists.

The hatred against the discoverers of a phenomenon which threatened to undo the cherished beliefs of science was as bitter and inexorable as that of the religious persecutors two centuries before. It was, in fact, of the same character.

A contemporary parallel to the disregard of the facts of hypnotism by science would seem to be the present attitude of science to extrasensory perception. I am not concerned here with the question whether this attitude is right or wrong, and I am not sure of it myself. I only want to show what I mean by scientific beliefs, the holding and applying of which is essential to the pursuit of scientific inquiry.

III

The positivist may admit that scientific interpretations include a fiduciary element, but will insist that even so there exists a core of hard facts or incontestable primary sensations which any theory will have to accept as such.

However, it is very difficult to discover any such primary sensations which are given previous to our interpretation of them. A child presented with a number of objects on a tray will notice only those with which it had some previous familiarity. The Fuegans, whom Charles Darwin visited from the *Beagle,* were excited by the sight of the small boats which took the landing party ashore, but failed to notice the ship itself, lying at anchor in front of them. Our eyeballs are full of small floating opaque bodies which we do not normally notice, but which fill us with alarm when some eye trouble calls our attention to them. There is a blind spot in our field of vision which can obliterate a man's head at a six-foot distance, but seems to have gone unnoticed throughout recorded history until comparatively recent times. To say that we have sensations which we do not notice seems hardly acceptable, for the moment we notice a thing, say by sight, we perceive it *as* something. We usually perceive it as being at some distance and as forming part of something else or standing out against other things as its background. Implicit in these perceptions will be the object's size and its being at rest or in motion. The perceived color of an object will largely depend on our interpretation of it. A dinner jacket in sunshine is seen as black and snow at dusk is seen as white, though the white snow sends less light into the eye than the black dinner jacket. Such facts as these leave little scope for sensations as primarily given data. It shows that even at the most elementary stages of cognition we are already committing ourselves to an act of interpretation.

There is always a measure of choice in our manner of perception, and whenever we see something in one way we cannot see it at the same time in a different way. A black spot on a white background may be seen either as a blot or as a hole, as the eye must choose between the two ways of seeing it. We may see a passing train at rest and feel ourselves moving, or the reverse, but we must choose between the two forms of perception. An attack on our senses may well compel our attention. But if it does, it will also compel perception and we shall commit ourselves to some way of receiving the impression and not know it in any other form.

These observations have an important general significance. When you adopt one way of looking at things you destroy at the same moment some alternative way of seeing it. This is the reason why disputation is deliberately used as a method of discovering the truth. In a courtroom, for example, counsels for the prosecution and for the defense are each required to take one side of the question at issue. It is supposed that only by committing themselves in opposite directions can they discover all that can be found in the favor of each side. If instead the judge should enter into friendly consultation with both counsels and seek to establish agreement between them, this would be considered a gross miscarriage of justice.

But it is not often realized that, even in the scientific handling of inanimate systems, different approaches are possible which are mutually exclusive. The laws of nature very often make definite predictions. For example, Boyle's law pv = const. is such a prediction of the changes of pressure accompanying the expansion or compression of a gas. Whether or not any particular gas under observation shall be judged to fulfill or falsify this prediction may still need to be decided; and then the theoretical prediction would be definite. Take, on the other hand, a radioactive atom which is prone to disintegration and of which we know the probable lifetime. Suppose this probable lifetime were an hour. It is quite easy to imagine an apparatus by which we could observe the decomposition of such a single atom–and to avoid irrelevant side issues we may imagine also that this atom is the only one of its kind in the world. Its probable life period would clearly predict something about the atom's behavior, but nothing so definite as pv = const. In accepting it to be true that the probable life period is an hour we commit ourselves to an expectation, and if it is not fulfilled–if the atom decomposes after five seconds or keeps us waiting for a week–we can only say that we are surprised; for our affirmation was only of the likelihood of an event and did not exclude the possibility that the unlikely would happen.

The two kinds of expectations which I have just described may be entertained in respect to the same situation, but they are mutually exclusive. We can say that the chance of throwing a double six with two dice is $1/36$; but we could not say this, nor anything about the chances of such a throw, if we knew exactly the mechanical conditions prevailing at the moment of the throw. We could predict from these the result of the throw–but the conception of chances would have vanished and would remain inconceivable for a system known in such detail. Thus a more detailed knowledge may completely destroy a pattern which can be envisaged only from a point of view excluding such knowledge.

Something very similar applies to a machine, the detailed observation of which may be wholly irrelevant and therefore misleading. What matters to the understanding of an object as a machine is exclusively the *principle* of its operation. The knowledge of such a principle, as defined for example by a patent, will leave the physical particulars of the machine widely indeterminate. The principle of the lever, for example, can be employed in such an infinite variety of forms that hardly any physical characteristic could apply to all of them. It represents a logical category which is always in danger of being obscured by a detailed description of an object to which it applies.

Again, there are inanimate objects which function as signs. Take, for example, marks on paper forming the letter "a." These marks, taken as a sign, must not be *observed* but *read.* Observation of a sign as an object destroys its significance as a sign. If you repeat the word "travel" 20 times in succession you become fully aware of the motion of your tongue and the sounds involved in saying "travel" and you dissolve the meaning of the word "travel."

Martin Buber and also J. H. Oldham have brought out the fundamental difference between treating a person as a person and as an object. In the former relation we *encounter* the person, in the latter we do not see it as a person at all. Love is a manner of encounter. A man or woman regarded in their purely physical aspects may be the object of desire, but cannot be truly loved. Their person has been destroyed.

The most important pair of mutually exclusive approaches to the same situation is formed by the alternative interpretations of human affairs in terms of causes and reasons. You can try to represent human actions completely in terms of their natural causes. This is in fact the program of positivism to which I have referred before. If you carry this out and regard the actions of men, including the expression of their convictions, as a set of responses to a given set of stimuli, then you obliterate any grounds on which the justification of those actions or convictions could be given or disputed. You can interpret, for example, this essay in terms of the *causes* which have determined my action of writing it down or you may ask for my *reasons* for saying what I do. But the two approaches—in terms of causes and reasons—mutually exclude each other.

IV

Positivism has made us regard human beliefs as arbitrary personal manifestations, which must be discarded if we are to achieve a proper scientific detachment. Belief must be rehabilitated from this discredit if it is henceforth to form a recognized part of our scientific convictions.

Scientific beliefs are not a personal concern. Even if a belief is held by one person alone—as may have been the case for Columbus's belief in a western route to the Indies when he first conceived it—that does not make it an individual preference like the love of one's wife and children. The beliefs of scientists concerning the nature of things are held with a claim to universal validity and thus possess normative character. I would describe science, therefore, as a normative belief, which I share; just as astrology is a normative belief which I reject—but is accepted by astrologers.

Turning now to the contention that beliefs are arbitrary, I shall have to enlarge somewhat on the holding of beliefs in general. Whoever embraces a belief accepts a commitment. Commitments are regularly entered upon not only by people who believe something, but by almost any living being, and particularly by all animals engaged in purposeful action. A floating amoeba will emit pseudopods in all directions, until its nucleus is left bare of protoplasm at the center. When one of the pseudopods touches solid ground, all the others are drawn in and the whole mass of protoplasm is sent flowing toward the new point of anchorage. Such is the amoeba's mode of locomotion. We have here the prototype of a phenomenon which is repeated in a million variants throughout the animal kingdom. There is coordination between the simultaneous movements of the animal's limbs and also between movements following upon each other in time. We may characterize such coordinated sequences by the fact that any part of the sequence is meaningless by itself, while each makes sense in conjunction with the other parts. Each can be understood only as part of a stratagem for the achievement of a result which, we have reason to believe, gives satisfaction to the animal, e.g., getting food or escaping from danger. The more roundabout are the methods employed in achieving a purpose, the more sagacious will appear their coordination and the more clearly will we recognize in them a sustained striving for that purpose.

To say that an action is purposeful is to admit that it may miscarry. If it is the purpose of animals to survive until they have reproduced themselves, then surely the vast majority of purposeful actions do miscarry, for only a small fraction of each generation of animals lives to beget young. In any case, no animal engaging in a purposeful action can be certain that the efforts it is about to invest will bear fruit. Nor can there be any certainty that an alternative course of action might not have had a greater chance of success. All purposeful action, therefore, commits its agent to certain risks. Purposeful forms of behavior are a string of irrevocable and uncertain commitments.

Commitments of this kind might be said to express a belief; where there is purposeful striving, there is belief in success. Certainly no one can be said truly to believe in anything unless he is prepared to commit himself on the strength of his belief. We conclude that the holding of a belief is a commitment of which human beings are capable, and which is a close analogy to the commitment in which animals universally engage when embarking on a purposeful course of behavior.

Let us now return to scientific beliefs in particular. When we say that an affirmation of a scientist is true or false, we usually have no need to rouse our fundamental scientific beliefs. We may turn our backs on them and take them for granted as the unconscious foundation of our judgment. But when some major question is at stake (like hypnotism, telepathy, etc.) our beliefs do become visibly active participants in the controversy, and we find it then more appropriate to say "I believe such and such to be true, or false." Such a belief may turn out to be true or false, as the case may be, but the affirmation of the belief falls into neither of these categories. The affirmation of a belief can only be said to be either *sincere* or *insincere.* Sincere beliefs are those to which we are committed, and a fiduciary commitment is, therefore, by definition sincere. Our commitments may turn out to have been *rash.* But it is in the nature of a belief that at the moment of its being held it cannot be fully justified, since it is in the nature of all commitments that at the time we engage upon them their outcome is still uncertain.

The only ground on which the sincere holding of a belief or the entering on any other kind of commitment can be criticized is for not having sufficiently taken into consideration its own possible rashness. But we must remember that any postponement of judgment for the sake of its reconsideration is itself a commitment. To go on hesitating for the sake of making more certain of one's decision may be the most disastrous, and indeed the most irresponsible, course to choose. I should say, therefore, that when a belief is both sincerely and responsibly held—that is, in conscientious awareness of its own conceivable fallibility—there is present a form of affirmation which cannot be criticized on any grounds whatsoever. It is a form of being the justification of which cannot be meaningfully questioned.

Such a situation is, of course, subject to revision, and the present moment's belief can be rejected or modified by the next moment's reflection, but this reflection, and its result, will be again an ultimate commitment, which so far cannot have yet become the object of reflection or criticism. But commitment must have duration. Any attempt to accompany it simultaneously by reflection is logically self-contradictory, and, if we persist, it results in the disintegration of our person. If we cannot loose ourselves at all, but feel compelled to observe ourselves in all we do, we become disembodied in the manner which Sartre has penetratingly described. People who cannot rid themselves of the feeling that they are "playacting" become incapable of any convinced action. The result is not a superior degree of detachment but an impotent nihilism.

We may conclude, in fact, that detachment in the rigorous sense of the word can be achieved only in a state of complete imbecility well below the normal animal's level.[2] In all states of mind above that, we are inevitably committed, and usually we are committed to an approach which excludes other approaches. The descriptive scientific approach as conceived by positivism is inadequate even for the handling of inanimate systems, in which we have to assess chances, or understand machines, or read signs; and when applied to persons–be they human or animal–and their actions, it dissolves them both as persons and as rational beings. This approach, far from representing a state of absolute detachment, is in fact a commitment to specific and, as it happens, extremely unreasonable presuppositions, to which no one would conceivably commit himself but for the fact that they are taken to provide the one completely detached, objective approach to the world.

Detachment in the ordinary and true sense always means commitment to a particular approach which we deem to be proper to the occasion, and disengagement from other points of view which for the time being are inadmissible. To hold the balance between

[2]I am thinking here of the behavior of decerebrated dogs and decorticated rats, and of the pure reflex behavior of incomplete lower organisms, such as Planaria, described by Kepner. In such cases we observe incoherent behavior, sustaining no purpose.

our alternative possible approaches is our ultimate commitment, the most fundamental.

V

The beliefs which men hold are mostly imparted to them by their early education. Others they acquire later through professional training and through the wide variety of educative influences which infiltrate our minds from the press, from works of fiction, and through other innumerable contacts. These beliefs form far-reaching systems, and though each of us is directly affected only by one limited part of them, we are committed by implication to the pattern of which this is a part.

This is obvious in the case of a scientist who is trained as a specialist from the start and, eventually, through his researches becomes an authority on a branch of his specialism; and who yet in this process absorbs the scientific outlook in all manifestations. The system of scientific beliefs is transmitted by such acceptances from generation to generation within the community of scientists. Systematic beliefs can be perpetuated only by large, highly differentiated communities.

The transmission of beliefs in society is mostly not by precept, but by example. To take science: there is no textbook which would even attempt to teach how to make discoveries, or even what evidence should be accepted in science as substantiating a claim to discovery. The whole practice of research and verification is transmitted by example and its standards are upheld by a continuous interplay with criticism within the scientific community. No one who has experienced the woeful unreliability of scientific output coming from places where scientific standards have not been firmly established by tradition, or who has felt the difficulty of doing good scientific work within such a milieu, will fail to appreciate the communal character of the premises on which modern scientific work is based.

Scientists are, of course, never unanimous on *all* questions. There may even be clashes from time to time about the general nature of things and the fundamental methods of science (as in the case of hypnotism, telepathy, etc.). Yet the consensus of scientific beliefs has not been seriously endangered during the past 300 years; not until the attempt by Soviet Russia to secede from the international community of science and establish a new scientific community, based on markedly different beliefs. Up to then, there was always between scientists in all parts of the world, and between each generation and the next, sufficient consensus of fundamental beliefs to assure the settlement of all differences.

The scientific community is held together and all its affairs are peacefully managed through its joint acceptance of the same fundamental scientific beliefs. These beliefs, therefore, may be said to form the constitution of the scientific community and to embody its ultimate sovereign general will. As each scientist is committed to these beliefs, the whole community may be said to be dedicated to them. The freedom of science consists in the right to pursue the exploration of these beliefs and to uphold, under their guidance, standards of the scientific community. For this purpose a measure of self-government is required, by virtue of which scientists will maintain a framework of institutions, granting independent positions to mature scientists, the candidates for these posts being selected under the direction of scientific opinion. Such is the autonomy of science in the West, which logically flows from the nature of the basic purpose and fundamental beliefs to which the community of scientists is dedicated.

The Marxist conception of science is different from that of the West, and its application in Russia has already led to serious changes in the position of science there and to a breach, at various points, between the scientific opinions of the East and West. The most far-reaching action in this direction was the official and sweeping repudiation of Mendel's laws, and of the whole conception of biology related to these laws, by the Soviet Academy on August 26th, 1948.

There was much indignant protest in Britain against this decision of the Soviet government and even more against the pressure exercised by the Soviet government, to which the Russian Academy had yielded in taking its action. I subscribe to these protests, but I wish their proper theoretical foundation were more clearly realized. If you protest in the name of freedom in general, it is embarrassing to admit that hitherto it was the anti-Mendelists and the whole school of Michurin and Lysenko whose publications were excluded from all the leading scientific journals and whose teachings were unrepresented in university curricula. Marxians were quite right in pointing out that there always exist accepted views on certain general issues of science which are imposed by scientific opinion on scientific journals, textbooks, and academic curricula, and from which candidates for scientific posts can dissent only at great peril to their future chances. They were right also in recalling that the views thus imposed were sometimes found later to be untrue and the dissenters vindicated.

We must accept it that the existing body of science—or at any rate its fundamental beliefs—is an orthodoxy in the West. Millions are spent annually on the cultivation and dissemination of science by the public authorities, who will not give a penny for the advancement of astrology or sorcery. In other words, our civilization is deeply committed to certain beliefs about the nature of things; beliefs which are different, for example, from those to which the early Egyptian or the Aztec civilizations were committed. It is for the cultivation of these particular beliefs—and these alone—that a certain group of people has been granted a measure of independence and official support in the West.

This is what we call academic freedom. Replace science as we know it by some other study we do not believe in and we cease to protest against political interference with its pursuit. Suppose, for example, that Lysenko and his supporters were given a clean 30 years to transform biology, physics, and chemistry in the image of dialectical materialism throughout the universities of the U.S.S.R.; and that subsequently, by some miracle, Marxism were abandoned by the government of the Soviet Union. We would certainly not uphold the academic liberties of the then occupants of scientific positions against an anti-Lysenko acting the way Lysenko does today, but this time for the re-establishment of our conception of science. We may demand a measure of freedom for almost any nonsense in a free country, but that is not what we mean by academic freedom.

Those who engage with Marxists in discussion about the freedom of science must face this situation. The Marxists are quite near the truth in saying that in demanding freedom we merely seek to establish our own orthodoxy. The only valid objection to this is that our fundamental beliefs are not just one orthodoxy; they are true beliefs which we are prepared to uphold. This true vision also happens to open greater scope for freedom than other, false visions; that is so, but in any case, our commitment to what we believe to be true stands first.

More generally, the freedom of science cannot be defended today on the basis of a positivist conception of science, which involves a positivist program for the ordering of society. Totalitarianism is a much truer embodiment of such a program than is the free society; as, indeed, consistent positivism must destroy the free society. A complete causal interpretation of man and human affairs disintegrates all rational grounds for men's convictions and actions. It leaves you with a picture of human affairs construed in terms of appetites checked only by fear. All you have to explain then in order to understand history, and with it politics, law, science, music, etc., is why at certain moments the appetite of one group gets the upper hand over its rivals. You have various options at this point. Marx and Engels decided the question in terms of class war. They affirmed that the class which, by taking control of the means of production, can make best use of them for the production of wealth, will prevail. When the time comes, the victory of the rising class is inevitable, though it can be achieved only by violence, for no ruling class can agree to its own annihilation. This theory was put forward as a scientific discovery, the discovery of the "laws of motion" governing society; and I believe that some conception of this kind does in fact inevitably follow from any consistent application of the positivist program to the affairs of man.

According to the positivist theory of society, no human judgment—be it in politics, law, or art, or any other field of human thought, including science itself—can be said to be valid except in the sense that it serves a certain power. In the Marxist version this is the power of the rising class, and today in particular the power of the vanguard of the rising class, as embodied in the Soviet government. That is the theory of science facing us in Russia today. Here the positivist movement, which had set out to establish the reign of science over all human thought, is culminating in the overthrow of science itself.

VI

The free society—of which a free scientific community naturally forms part—can be defended only by expressly recognizing the characteristic beliefs which are held in common by such a society and professing that these beliefs are true. The principal belief—or, shall I say, the main truth—underlying a free society is that man is amenable to reason and susceptible to the claims of his conscience. By reason is meant such things as the ordinary practice of objectivity in establishing facts and of fairness in passing judgment in individual cases. The citizens of a free society believe that by such methods they will be able to resolve jointly—to the sufficient satisfaction of all—whatever dissension may exist among them today or may arise in the future. They see an inexhaustible scope for the better adjustment of social institutions and are resolved to achieve this peacefully, by agreement.

Just as on a smaller scale the scientific community organizes, disciplines, and defends the cultivation of certain beliefs held by its members, so the free society as a whole is sustained for the practice and by the practice of certain wider, but still quite distinctive, beliefs. The ideal of a free society is in the first place to be a *good* society; a body of men who respect truth, have a passion for justice, and love their neighbors. It is only because these aspirations coincide with the claims of our own conscience that the institutions which secure their pursuit are recognized by us as the safeguards of our freedom. It is misleading to describe a society thus constituted, which is an instrument of our consciences, as established for the sake of our individual selves; for it protects our conscience from our own greed, ambition, etc., as much as it protects it against corruption by others. Morally men live by what they sacrifice to their conscience; therefore the citizen of a free society, much of whose moral life is organized through his civic contacts, largely depends on society for his moral existence. His social responsibilities provide the occasion for a moral life from which men not living in freedom are debarred. That is why the free society is a true end in itself, which may rightly demand the services of its members in upholding its institutions and defending them.

The fiduciary formulation and acceptance of science fits in with our fiduciary conception of the free society. Scientific beliefs are a part of the beliefs cultivated in such a society and accepted by its members. That is their valid defense against Marxism. But we must realize that this defense accepts a position of knowledge in society which in many ways recalls that assigned to it by Marxism. It implies that the free society upholds an orthodoxy which excludes certain suppositions that are widely current today. Any representation of man and of the affairs of man which, if consistently upheld, would destroy the constitutive beliefs of a free society, must be denied by its orthodoxy. A behaviorism which denies the very existence of the moral sphere for the sake of which the free society is constituted, or a psychology which discredits as mere secondary rationalization the purposes which a free society regards as its mainsprings, will be rejected by its orthodoxy.

A free society would cease to exist if its members ever admitted that some major conflict had to be settled by sheer force within the society to which they belonged. Such an admission would therefore be subversive of the free society and constitute an act of disloyalty to it. Nor should members of a free society ever admit that experience can disprove that moral forces operate in history, any more than a scientist will admit that experience can disprove the scientific conception of things. They should, on the contrary, insist on searching history for the manifestation of a sense of justice, and try to discover in every reconciliation and pacification the fruits of human confidence responding to human confidence.

Science or scholarship can never be more than an affirmation of the things we believe in. These beliefs will, by their very nature, be of a normative character–that is to say, claiming universal validity–and they must also be responsible beliefs, held in due consideration of evidence and of the fallibility of all beliefs; but eventually they are ultimate commitments, issued under our personal judgment. To all further critical scruples we must at some point finally reply: For I believe so.

We are entering in this century into a period requiring great readjustments. One of these is to learn once more to hold beliefs. Our own beliefs. The task is formidable, for we have been taught for centuries to hold as a belief only the residue which no doubt will conceivably assail. There is no such residue left today, and that is why the ability to believe with open eyes must once more be systematically restored.

5

SCIENTIFIC BELIEFS

I

It is widely assumed today that the pursuit of science represents a field of intellectual activities which does not require the acceptance of any doubtful beliefs. Science is taken to be, by its very nature, positive and demonstrable, differing in this respect not only from religious belief but also from any ethical conviction, such as a belief in justice or any other moral standard to which we are committed.

As early as John Locke a distinction of this kind was drawn in respect to religious truths. These, Locke says (1692, p. 144), are not capable of demonstration, "how well-grounded and great soever the assurance of faith may be wherewith it is received; but faith it is still and not knowledge; persuasion and not certainty." The two and a half centuries of scientific triumphs that have passed since Locke drew this distinction between faith and demonstrable knowledge have greatly added to the prestige of science as the embodiment of knowledge that is unambiguous and objective. A passionate affirmation of what some scientists believe science to be was given in recent years by the distinguished American psychologist, Clark L. Hull (1943), in his *Principles of Behavior.* The essence of scientific objectivity lies, he says, in establishing rigorous mathematical relations between measured variables. Given the values of one set of variables, science predicts

First published in *Ethics,* 61 (Oct.):27-37, 1950.

exactly the values of another set. A genuine scientific theory must operate like a calculating machine, which, once the keys representing the dividend and the divisor have been depressed, determines the result automatically (p. 24).

Such testimonies of scientists to the exactitude and objectivity of science carry conviction far and wide today. Indeed, I would hardly venture to oppose them here but for the recent occurrence of a massive event that is very difficult to reconcile with this conception of science. It has happened that "a great scientific nation has repudiated certain basic elements of scientific method and in so doing has repudiated the universal and supranational character of science." The words are by Julian Huxley, writing on Soviet genetics in *Nature* (1949b), and I think that they give a correct description of the schism between the scientific opinions of the West and the views of scientists under communism. They show that certain substantial parts of science, which carry conviction and are accordingly recognized as science in one part of the world, fail to convince and are contemptuously rejected in another part. This seems to call seriously into question the claim that science is rigorous and impersonal and suggests that there are certain not indubitable beliefs involved in our acceptance of science–of our science–as valid.

With this in mind, let me point out the very substantial flaws which the rigorously positive conception of science contains and show that it needs to be supplemented by fiducial elements–which I shall call "scientific beliefs"–for a true picture of science.

It is easy to demonstrate that the positivistic model of science–so emphatically endorsed, for example, by Hull in the passage which I quoted–as a set of empirically established mathematical relations between measured variables, is incomplete, so incomplete as to be no more than a caricature of science. Suppose that the evidence on which a scientific proposition is to be based consists of a number of measurements made at various observed times, or else in coincidence with a series of observations made on some other measurable parameter. Let us, in other words, have pairs of two measured variables X and Y. Can we ever decide from a series of points of X plotted against Y whether there exists a functional relationship $X = f(Y)$ and, if so, what it is? Clearly, we can do nothing of the kind. Any set of pairs of X and Y values is compatible with an infinite number of functional relations between which there is nothing to choose from the point of view of the underlying data. To choose any of the infinite possible functions and give it the distinction of a scientific proposition is so far without any justification. The measured data are insufficient for the construction of a definite function $x = f(y)$ in exactly the same sense as two elements of a triangle are insufficient to determine a particular triangle.

It follows that, if scientific propositions were equivalent to mathematical functions of one exactly measurable variable depending on another exactly measurable variable, the chance of discovering any such proposition from the observation of the variable in question would be exactly zero. These propositions would be strictly underivable from the evidence to which they refer. Moreover, if they were, nevertheless, given in some inconceivable manner, these propositions would be strictly unverifiable. Successful prediction would not by itself achieve a finite measure of confirmation for the proposition in question. It would only add a number of observations, the predicted observations, to the previously given series of measurements and could not change the fact that any series of measurements is incapable of determining any particular function holding between the measured variables (see Polanyi, 1946, pp. 7-8).

Since some of you may be shocked by this conclusion (even though it merely restates what Hume said long ago), I shall illustrate it a little further. Suppose a player of roulette observes the number of reds and of blacks that turned up in a hundred consecutive throws. He may plot them in a graph and derive a function in the light of which he will make a prediction. He may try it out and win. He may try it again and win. And win a third time. Would that prove his generalization? No; in our view it would only prove that some roulette players are very lucky–i.e., we would consider the fulfillment of his predictions as mere coincidence.

A few years ago there appeared in *Nature* a table of figures proving with great accuracy that the time of gestation, measured in days, of a number of animals ranging from rabbits to cows is a multiple of the number π (see Table 1). Yet an exact relationship of this kind makes no impression on the modern scientist, and no amount of further confirmatory evidence would convince him that there is any relation between the period of gestation of animals and multiples of the number π.

It is, of course, not inconceivable that in one hundred years' time people may think differently on this point. About one hundred years ago the great Helmholtz declared: "Neither the testimony of all the Fellows of the Royal Society, nor even the evidence of my own senses could lead me to believe in the transmission of thoughts from one person to another, independently of the recognized channels of sensation" (quoted by Ehrenwald, 1948, pp. 23-24). I think that scientists would no longer unanimously concur with the decision of Helmholtz to reject in advance all evidence of extrasensory perception as fallacious, though very likely most of them would.

TABLE 1

AVERAGE GESTATION PERIOD AND $n\pi$*

n	$n\pi$	Average Gestation Period (Days)	No. of Pregnancies	Animal
10...	31.416	31.41	64	English rabbit
36...	113.097	113.1±0.12	203	Pig
48...	150.796	150.8±0.13	195	Karakul sheep
		150.8±0.19	391	Black Forest goat
49...	153.938	154	?	Saanen goat
92...	289.026	288.9	428	Simmental cow

*J. H. Kenneth (1940, p. 620).

Not long ago listeners to the Third Programme in England were given a jolt by a broadcast discussion on astrology, in which an astrologist produced an array of carefully sifted evidence of astrological predictions that had come true. Possibly in another thousand years we shall be inclined to take such evidence seriously. Today we are not; and the fact remains that at all times there are a great many conceivable empirical relations that are rejected as absurd from the start and for which empirical evidence, therefore, is not accepted as valid, while evidence of a similar amplitude and apparent accuracy would be accepted as more than adequate to prove a relationship against which no such initial objection had been made.

This illustrates and substantiates the conclusion derived previously by the logical analysis of the positivistic model of science, namely, that the fulfillment of predictions in terms of observations is not in itself capable of validating a scientific statement. And we may add that even the converse of this is true. Our general conceptions of the nature of things cannot be strictly contradicted by experience, for they can always be expanded so as to cover any experience. This is often true even of specific scientific theories. We may quote as an example the Copernican theory of cycles and epicycles, which is capable of covering, to any required measure of approximation, any conceivable planetary motion. Or take Dalton's law of simple chemical proportions or Haüy's law of rational indices in crystallography. No observed chemical compositions of two compounds can flatly contradict Dalton's law, for its prediction is not exactly determinate, and the same is true of the law of rational indices in crystallography.

This is not to say that a theory cannot *sometimes* be flatly contradicted by observation. But the current positivistic story that a scientist immediately drops a hypothesis the moment it conflicts with experience is a pure myth. No true scientist acts in this clumsy manner. Niels Bohr did not drop his theory of spectra, which was confirmed only by one single type of atom—that of hydrogen—and broke down at the very next step, when applied to helium. The periodic system of elements was not abandoned when two pairs of elements could be fitted into it only in the reverse sequence of their atomic weights. Chemistry held on firmly to the cyclic formula of benzene proposed by Kekule in 1859, even though it became clear, as the years passed by, that the two different disubstituted derivatives which it postulated did not, in fact, exist. Scientists will often tolerate such contradictions to their theory, regarding them as anomalies which may be eliminated in the course of time by an amplification of the theory. Whether they should abandon a theory or not in any particular case can be determined by no fixed rule. The scientist's decision depends on the strength of the beliefs in the light of which he interprets his observations, and we approve of this decision if we share these beliefs.

Those who are convinced that science can be based exclusively on data of experience have tried to avoid the weight of such critical analysis by reducing the claims of science to a more moderate level. They point out that scientific propositions do not make claim to certainty but claim only to be likely; that they make only provisional statements, which are always subject to revision.

These reservations are, however, beside the point. If anyone claims that, given two angles, he can construct the triangle, his claim is equally nonsensical whether he claims to give a true construction or merely a probable construction or the construction of a merely probable triangle. The selection of one element out of an infinite set of elements all of which satisfy the conditions set by the problem remains equally unjustifiable whatever positive quality we attach to our selection. Its value is exactly nought. In fact, scientists would object just as much to serial rules in games of chance or to astrological predictions or to relations between the time of gestation of animals and the number π, whether these are claimed with certainty, only with probability, or even merely provisionally. They would be regarded as no less nonsensical for that.

Such efforts to minimize the claims of science in order to make scientific discovery less puzzling remind one of the debate about the actual distance which the martyr St. Denis covered away from Montmartre, carrying his head in his arms. Mme. du Deffand settled this neatly by exclaiming, "But surely, in such a situation *ce n'est que le premier pas qui coute!*" ("it is only the first step that counts").

Other attempts to lessen the burden of responsibility on the scientists' shoulders do not prove more successful. Science, it is urged, does not claim to discover the truth but only to give a description or summary of observational data. But why then object to astrology or to the description of periods of pregnancy in multiples of the number π? Surely for no other reason than that they are not held to be *true* or *rational* descriptions, which brings the problem back exactly to our starting point.

It has been suggested that scientists are giving the simplest description of their observations. But this is blatantly false. Scientists do not reject astrology, magic, the coincidence of pregnancy periods with multiples of the number π, or extrasensory perception because these are not simple descriptions of the observed facts. And I would even say that human ingenuity could devise no more *involved* description than that given by the general theory of relativity, of the facts to which this theory refers. The adjective "simple" can be used as a distinctive mark of scientific statements only if it is tortured into meaning "rational" and finally made to coincide with "true," which brings us back again to where we stood before.

II

The fact that serious and wise people with penetrating minds have so long subscribed to such rigmaroles about the nature of science can be understood only as expressing a deep, underlying urge of our modern civilization. It is due to a fundamental reluctance to recognize our higher faculties, which our empiricist philosophy cannot account for. We dread to be caught believing–and, in fact, knowing–things which are not demonstrable by the measurement of observed variables. So we fabricate all kinds of pretenses and excuses and describe our most profound insights as merely "economic descriptions" and speak of our most assured convictions as mere "working hypotheses." This serves us as a verbal screen behind which to hide our philosophically unaccountable power of discovering the truth about nature and our wholehearted commitment to the truths which we have so obscurely acquired.

The positivistic movement is, of course, only part of the broad trend engendered by modern empiricism, which induces us to camouflage in utilitarian colors our transcendent faculties and obligations, in order that they may pass muster before a skeptical philosophy. Like clowns imitating puppets, we pretend to be pulled by strings, so as to conform with a mechanistic conception of man. It is part of this pattern that we dare not confess that we hold the scientific beliefs which we actually hold, for fear of the empiricist policeman behind us. We look carefully over our shoulders and pick our words appropriately, to avoid saying anything so metaphysical as that science inquires into the nature of things or that it seeks to explain them–for fear of offending the ruling assumptions about the strictly mechanical origin of science.

Yet this is precisely the clue which we must pursue in order to replace the positivistic model by a more adequate reference to science. We must openly declare the beliefs which we hold about the nature of things which guide our pursuit of science. We must admit that these beliefs are acquired by us uncritically in the course of our early education or subsequent apprenticeship to science and that no succeeding critical examination of these basic beliefs can ever eliminate all fiducial elements of our thinking. Let me turn to the exposition of this situation.

Our most general belief, which underlies all our scientific thinking, is the naturalistic view of the universe that is current among us today. When a scientist rejects *in limine* any empirical evidence for the reliability of predictions from horoscopes, for the potency of magic or the powers of witchcraft, he acts in the light of the naturalistic conception of things inculcated in us by our earliest upbringing. We do not take to this enlightened view of nature as a matter of course, without any pressure from outside. On the contrary, there seems to exist a strong natural bias toward a magical view of the world in which all events that affect us appear to embody an intention on the part of personal forces. As small children we encounter the objects of our surroundings indiscriminately as personal agencies. The I-Thou relationship which we establish with persons also prevails in our intercourse with impersonal objects. Only later do we class these objects in an I-It relationship and recognize the distinction between persons and things. All works of fiction and all but the driest chronicles of history embody important elements of the magical outlook. In an epic poem or in a novel nothing happens to a character that is not meaningful in the context of the story. Accidental death in a novel is never truly accidental, that is to say, truly meaningless, but is an event fitting into the context of values to which the author bears witness. The magical outlook is similar to that of the child, in that it lacks the clear subject-object relation of modern man; and it resembles the world of fiction in the assumption that anything that happens to man is part of a meaningful story. Modern education reduces our natural predilections which favor the magical outlook and inculcates in us conceptions which primitive people do not possess. Thus we acquire a clear subject-object relationship and the idea of natural causes, a corollary to which is the conception of accident. Lévy-Bruhl has shown that this supposition forms the logical basis of the naturalistic outlook, according to which events affecting human fate may be purely accidental, i.e., entirely meaningless.

III

We acquire our naturalistic system of explanations as we first acquire speech, uncritically absorbing the idiom of our elders. We do not learn our mother tongue in preference to a hundred other living languages, as well as Esperanto, Ido, ancient Etruscan, and the language of the *Principia Mathematica*, after examining the vocabularies and grammars of each and testing their relative appropriateness. We are born into a language, and we are also born into a set of beliefs about the nature of things. And, once brought up in our beliefs, we embrace them with sufficient conviction to participate in imposing them on the next generation. We wager our lives on these beliefs, and we share in building on these beliefs the life of the whole community which holds them. Indeed, this community is largely constituted by this joint commitment. It is dedicated to the pursuit of a technology and a science, of a system of jurisdiction and of political discussion, and of a way of life in general that is profoundly affected by the holding of these constitutive beliefs.

People brought up in a system of magic, witchcraft, and oracles are equally committed to their beliefs. Evidence which in our eyes would blatantly expose the worthlessness of their oracles does not in the least embarrass such people. This was made abundantly clear by Evans-Pritchard (1937) in his work on the Central African tribe of Azande. "Azande," he says, "observe the action of the oracle as we observe it, but their observations are always subordinated to their beliefs and are incorporated into their beliefs and made to explain them and justify them. Let the reader consider any argument that would utterly demolish all Zande claims for the power of the oracle. If it were translated into Zande modes of thought it would support their entire structure of beliefs" (p. 339). Their blindness (as we would regard it), far from being due to stupidity, is sustained by a remarkable ingenuity. "They reason excellently," writes Evans-Pritchard, "in the idioms of their beliefs, but they cannot reason outside or against their beliefs because they have no other idiom in which to express their thoughts" (p. 338).

It is unquestionable that if you or I had been brought up among Azande, we would think as they do and be impervious to the evidence which now convinces us of the foolishness of their beliefs. This inevitably raises the stock argument used by religious doubters for centuries, which points out that beliefs accepted uncritically on the strength of a local authority will always testify to their own validity but that such testimony is worthless and that therefore such beliefs are unfounded. Yet we do not accept this argument in respect to the different beliefs which Azande and ourselves hold with regard to the nature of things. At any rate, I am prepared to congratulate myself on not having been brought up in a system of beliefs such as those held by Azande, which I would otherwise falsely believe in. This is admittedly like saying that I am glad I dislike spinach, as otherwise I would eat it, though I hate it. When we pat ourselves on the back for being more enlightened than Azande, we rely on the same authority to which we blindly submitted in acquiring the supposed enlightenment. The witness which we bear to this authority is part of its own teaching. And this is quite proper. The regional diversity of beliefs and their social rootedness can shake our beliefs only if these are already losing ground for other reasons.

It is indeed logically impossible for the human mind to divest itself of all uncritically acquired foundations. For our minds cannot unfold at all except by embracing a definite idiom of beliefs, which will determine the scope of our entire subsequent fiducial development. As Evans-Pritchard observes, even the doubters among Azande express their dissent in alternative Zande beliefs. This is not to say that our beliefs are immutable. They are constantly remolded in the course of being applied to new objects. I shall illustrate in a moment how great pioneers, like Pythagoras or Galileo, opened up new vistas of scientific beliefs which guided subsequent periods of research. I myself believe that even our commonly accepted naturalistic I-It relation may need revision in its application to man, so as to avoid impairing our appreciation of man as a responsible moral being. But we must not fall into Zeno's fallacy of assuming that something that is in flux or in motion is therefore not at some particular place at every moment. There are no limits set to the process of critical reflection, including, of course, our taking into account what people believe elsewhere and what we ourselves may think in the future. But, for all that, there is always a set of beliefs to which we are committed at any particular moment. And this commitment is, of necessity, largely determined by the first uncritical acceptance of our education in early life. In our own case, the case of modern man, we are committed in this manner to a naturalistic conception of the universe which is the major unrebuttable premise of all modern natural science.

Such disclosures of the fiducial foundations of science may be annoying to those brought up to regard science as universal and strictly objective, as an aggregate of $x = f(y)$ functions vigorously derived from measured data. They may feel uneasy at being deprived of their ideal of science in the image of Hull's calculating machine, clicking out strictly impersonal responses. I should like to offer them a reply to their apprehensions, even though by so doing I may momentarily add to their vexation, by evoking here the shades of St. Paul. I would suggest that, instead of analyzing the powers of science as something operating outside us, we try to clarify the obligations incumbent on a scientist, since that is what we are concerned with and, indeed, that is all we have to do. The simplest expression that I know of the scientist's obligation can be stated in terms of the Christian paradox, that man is called upon to try the impossible but is not expected to achieve it. As scientists we must seek a truth which is unambiguous and universal, even though at the same time we must recognize that this is impossible and, indeed, strictly speaking, meaningless. In science, just as in the jury box or in the voting booth or in the recruiting office, you must commit yourself on grounds which, on reflection, must necessarily appear deficient. In science, as elsewhere, we must choose, within the limits of time and other given circumstances, whether to affirm or to refrain from affirming and what to say and how to say it. Whatever our decision, it will fulfill our whole obligation if we have previously exhausted, within the limits imposed by the situation, the requirements of our ideal as interpreted by our conscience. All that can be required of us is that we should seek the universal in the light of such guidance as we possess. It is *not* required of us that we should decide our problems on the supposition that we were born in no particular place, in no particular time, and endowed with no personal judgment of our own. The fellow in the old joke who tells you, when you ask your way, that "he would not start from there" is talking logical nonsense. Our duty is always here and now; and mercifully that is all that is demanded of us. If we had been born among Azande, we would have Zande problems; as it is, we have our own, which include—as they do not include for Azande—the pursuit of science.

But perhaps I had better resume the exposition of the fiducial foundations of science so as to have an ampler range of reference on which to base my analysis. Consider our rejection of evidence purporting to show that the period of gestation of animals measured in terrestrial days is a multiple of the number π. This rejection does not follow from our naturalistic outlook and, indeed, expresses a comparatively recent belief of science. I should say that a scientist like Kepler would by no means have regarded as absurd the relationship suggested here. He had himself derived the existence of the seven planets and the relative sizes of their orbits by speculations of a similar kind. Take the five regular solids, or "perfect solids," as they used to be called—the tetrahedron, the cube, the octahedron, etc.—constructing a sample of each with the same length of edge. They form a series of increasing size, and you can imagine each polyhedron contained in the next. Then draw the inscribed and enveloping spheres, and you obtain, according to Kepler's theory, the orbits of the six planets recognized in Kepler's time. Today this suggestion appears absurd to scientists, though there are many people regarded by scientists as cranks who adhere to the older scientific belief, going back to Pythagoras, that nature is governed by number rules and standards of geometrical perfection. We should bear in mind that the revival of science by Copernicus was based on Pythagorean suppositions. To Copernicus, as to Pythagoras twenty-one hundred years before him, the problem of astronomy was defined as the explanation of the heavenly events on the assumption of circular and uniform motion.

It was only with Galileo and Gassendi, at the opening of the seventeenth century, that mathematical relationships were reduced to mere expressions of mechanical laws, while the presuppositions of science concerning the nature of things underwent a transformation on the lines of the ancient Democritean atomistics. In this view the real nature of things does not consist in numbers but is constituted by matter in motion. This conception of the universe as an aggregate of material particles which in themselves are colorless and tasteless and neither hot or cold–just as needles are in themselves neither pricking nor stitching–set out a new program for science which still largely dominates the popular imagination today. The ideal pursued by this program was defined in the eighteenth century by Laplace in his *Mèchanique Cèleste:* "An intelligence knowing at a given instant of time all forces acting in nature as well as the momentary positions of all things... would be able to comprehend the motions of the largest bodies of the universe and those of the lightest atoms in one single formula, provided his intellect were powerful enough to subject all data to analysis; to him nothing would be uncertain, both past and future would be present to his eyes."

What Laplace says here by no means sounds so absurd today as do Pythagorean teachings; yet, actually, Laplace's belief is no longer accepted in science. The assumption that only extension, mass, and motion are primary properties of matter became an obstacle to scientific progress as far back as Faraday's investigation of electricity and magnetism. The discovery of the electron was long delayed on account of the reluctance to admit that electrical properties should be regarded as ultimate qualities, not reducible–like heat, noise, smell, etc.–to mass in motion. The assumptions of Galileo and Laplace led also to a mass of futile speculations about the mechanical properties of the electromagnetic ether. And, finally, it turned out that the conception of Newtonian space as a framework at absolute rest, which is implicit in Laplace's vision, is absurd and must be discarded.

The recklessness of Laplace's teaching is today fully exposed to critical view, yet its convincing power is still alive. A contemplation of this skin which we have only just shed may help us to realize the nature of the commitment in which we are involved today in accepting science as valid. It reveals the two aspects of our actual scientific beliefs. Every belief is both a free gift and the payment of a tribute exacted from us. It is given on the personal responsibility of the believer, yet in the clear assumption that he cannot do otherwise. The two poles of a fiducial commitment, the personal and the universal, thus become apparent, and we can see that they are logically complementary and inseparable.

Hence we can now discern the fundamental fallacy of the positivist model of science. It tries to construct a machine which will produce universally valid results. But universal validity is a conception which does not apply outside the commitment situation. Any reference to it is merely a manner of expressing our submission to an ultimate obligation and can appear only as part of a fiducial declaration. The attempt to construct something universally valid, prior to any belief, is logically nonsensical.

Science can never be more than an affirmation of certain things we believe in. These beliefs must be adopted responsibly, with due consideration of the evidence and with a view to universal validity. But eventually they are ultimate commitments, issued under the seal of our personal judgment. At some point we shall find ourselves with no other answer to queries than to say, "Because I believe so." That is what no set of rules, or any model of science based on a system of rules, can do; it cannot say, "Because I believe so." Only a person can believe something, and only I myself can hold my own beliefs. For the holding of these I must bear the ultimate responsibility; it is futile, and I think also ignoble, to hunt for systems and machines which will take that burden from me. And we, as a community, must also face the fact that there is no system of necessary rules which will relieve us from the responsibility of holding the constitutive beliefs of our group, of teaching them to the next generation and defending their continued profession against those who would suppress them.

We can now more readily imagine that other people do not share these beliefs and are prepared to commit themselves to a somewhat different fiducial system. For example, J. B. S. Haldane (1949) wrote in connection with the Soviet genetics controversy that, if genes were unchangeable, he, as a Marxist, could not believe in them. Thus Haldane's anterior acceptance of Marxism determines to some extent his scientific beliefs. There is little doubt that Lysenko, Prezent, and a great number of their followers passionately believe that the Mendelian laws of inheritance are false, because in their view they contradict dialectical materialism and what they call "the revolutionary principle of changing nature for the benefit of the people" (*Proceedings of the Presidium of the*

U.S.S.R. Academy of Science, Aug. 26, 1948, quoted by Huxley, 1949a, p. 39).

There is also among Soviet thinkers a smoldering hostility to modern physics, which seriously flared up at the beginning of 1949 but has so far not led to a showdown (Ashby, 1950). It demands a return to the materialistic beliefs of Galileo-Gassendi-Laplace which were abandoned in the formulation of relativity and quantum mechanics. This materialistic conception, which has dominated men's minds for three centuries and was held to be self-evident by critical philosophers of the caliber of Kant, is, of course, still strongly rooted in popular beliefs. Soviet philosophers may well find grounds for adopting it once more, even if this causes difficulties in modern science. We may recall that two great German physicists, both Nobel Laureates, Philip Lanard and Johannes Stark, adopted the same position less than 15 years ago.

I think we should admit, therefore, that there is a genuine divergence of belief as to the nature of things between an important group of communists both inside and outside Russia and almost all the noncommunist scientists in the West. Naturally, the communists in question are trying to gain general acceptance for their fundamental premises by incorporating them in the curriculums of schools and universities. They try to displace everywhere scientists who in their view teach nonsense and pursue absurd lines of research by others who share their own fundamental beliefs.

We know that the process is speeded up by hateful forms of terror. But it would be equally destructive to the freedom of science if all the relevant decisions had been taken by democratic methods.

Apart from our detestation of their brutality, our objections to the intervention of the Soviet authorities in scientific life are all rooted in our disbelief of the new premises of science which they accept. There is hardly anything that Lysenko has done which would not be pardonable if it had been done to suppress some form of quackery that had become rampant in scientific institutions, so as to make way, instead, for the true methods of science. Ultimately, therefore, our protests must rest on the affirmation that our own scientific beliefs are true. There is no way of divesting ourselves of the responsibility for holding these beliefs and for committing ourselves to their dissemination and defense.

6

ON THE INTRODUCTION OF SCIENCE INTO MORAL SUBJECTS

I

When David Hume, hoping to emulate Newton's method, inscribed his first published work as *A Treatise of Human Nature, Being an Attempt to Introduce the Experimental Method of Reasoning into Moral Subjects,* he broached a program which he never fulfilled. He proceeded to study man in the light of common experience and used the example of science merely for letting reason operate unrestrained by religious faith or any other traditional beliefs that he wished to oppose. No determined movement for the application of scientific methods to the understanding of man and society was undertaken on a broad front before the opening of the present century. Yet today this movement has gathered such powers that it is creating a general tension throughout our culture, similar to that which at an earlier time the rebellion of reason against religion aroused, but even more comprehensive in its scope.

Look at some of the forms taken on by the modern scientific method when applied to human concerns. One of these is behaviorism, of which there are many shades. Radical behaviorism proposes to represent all human actions by constructing a robot which could perform all these actions and indeed appear to live the whole mental life of man, without the presence of any sentience in it. What is moving man becomes an irrelevant byplay.

A strict behaviorism is the logical terminus of looking at man in a completely detached manner, in accordance with the accepted

First published in *Cambridge Journal,* 7 (Jan.): 195-207, 1954.

ideal of the scientific method. The same manner of approach has therefore also been applied to the social functions of man. There is a school of thought in jurisprudence which defines law as what the courts will do, and which proposes accordingly to transform the study of law into a scientific observation of the way law courts behave in response to cases brought before them. Such legal behaviorism leaves out of the law the function of guiding the judge as to how he ought to decide a case and accordingly deprives the judge of any grounds on which he could seek guidance in his efforts to reach the right decision.

A tribe, a nation, or other social group may be described in purely objective terms with similar results. Most anthropologists insist on carrying out their analysis of society without mention of good and evil. Social life is thus found to rest on institutions which fulfill certain functions for the maintenance of society in its existing form. And this being all that the anthropologist is allowed to say, the terms by which he will describe the achievement of the noblest function in society will apply equally to its vilest aberrations. We find a distinguished anthropologist representing such practices as the unspeakably cruel murder of supposed witches, as a cultural achievement. "Some social systems are much more efficient than others in directing aggression into oblique or non-disruptive channels. There is no doubt that witchcraft is Navaho culture's principal answer to the problem that every society faces. How to satisfy hate and still keep the core of society solid" (Kluckholn and Leighton, 1946, p. 177). Anthropologists have similarly described head-hunting as fulfilling a social function in the societies in which it is practiced. "The religion of the Eddystone Islanders," writes Gordon Childe, "provided a motive for living and kept an economic system functioning" (1942, p. 15). Head-hunting, which formed part of Eddystone culture, proved wrong in his view only because, by keeping down numbers, it made improvement in material equipment superfluous and eventually left the islanders a prey to British conquerors.

For this kind of scientific anthropology, social stability is the only accepted value and therefore becomes the supreme social value. Yet all the time we know (and the anthropologist of course knows it too) that the stability of evil is the worst of evils.

So much for the anomalies that arise from examining man and human affairs from outside in the detached manner which is currently accepted as the ideal of the scientific method. Another set of similar tensions is generated by any attempt to judge our own mental activities according to the criteria which scientific statements must fulfill, or which they are, at any rate, currently believed to fulfill. For the past 50 years it has been hammered into us with never ceasing vigor that science is concerned only with verifiable (or at least falsifiable) statements and must be purified of all other elements. The Viennese school of philosophy has generalized this principle into a universal critique of human utterances. It has pointed out, for example, that if you say that it is wrong to bear false witness, you have made a statement which cannot be proved or disproved by the facts. No conceivable occurrence, no measurement or observation, can decide the question whether any action is moral or immoral, just or unjust, good or evil. Hence to call something immoral, unjust, or evil has no verifiable meaning. And it appears doubtful then whether it has any meaning at all beyond that of an exclamation, such as one may utter when biting into a worm inside an apple or when shouting "Boo!" to stop others from doing things which you find disgusting, or which for any reason you want to prevent from happening.

II

This conception of moral judgments is clearly unsatisfactory. When we express moral condemnation or approval, or when we seek guidance in a moral dilemma, there is always a reference to moral standards which are assumed to be given to us, in a general form. Moreover, while we use these standards for criticizing others, or for seeking moral guidance for our own problems, we hold them uncritically, as a matter of faith. To act on such faith may involve a great sacrifice, as when Socrates refused to flee from prison to escape execution, and such action seems quite different from the emitting of groans or snarls. On the other hand, our uncritical allegiance to such unsubstantial things as our moral standards seems to clash altogether with the currently accepted canons of scientific thought.

These perplexities have not been overlooked, but the remedy which has been offered for them seems to me inadequate. It consists in suggesting that science deals only with the factual aspect of things, while their value can be appreciated only by other modes of thought, such as we use in moral judgment. I believe that this division is false in principle and therefore impracticable. I shall argue that on the one hand it sets for science an ideal of objectivity which would discredit large and perfectly sound parts of science and which, indeed, if strictly applied, would invalidate all empirical science; and that on the other hand it deprives our valuations of the support which they may justifiably derive from their continuity with similar acts within science. I shall suggest that we should try to mend the break between science and our understanding of ourselves as sentient and responsible beings by incorporating into our conception of scientific knowledge the part which we ourselves necessarily contribute to such knowledge.

I shall introduce this revision of our conception of science by analyzing some anomalies which arise within science itself from the currently accepted ideal of the absolute objectivity of science.

Let me recall the terms in which Laplace, writing in 1795, defined a perfect scientific knowledge of the universe. If we knew at one moment of time, Laplace affirmed, the exact positions and velocities of every particle of matter in the universe, as well as the forces acting between these particles, we could compute by the laws of mechanics the positions and velocities of the same particles at any other date, whether past or future. To a mind thus equipped, he said, all things to come and all things gone by would be equally revealed. Such is the complete knowledge of the universe as conceived by Laplace.

In examining this ideal of universal knowledge I shall pass over the fact that it would have to be transposed into quantum mechanical terms today, for this would only introduce complications without affecting the grounds of my criticism. My main criticism of the kind of universal knowledge defined by Laplace is that it would tell us absolutely nothing that we are interested in. Take any question to which you want to know the answer. For example, having planted some primroses today, you should like to know whether they will bear blossoms next spring. This question cannot be adequately answered by a list of atomic positions and velocities at some future moment. It must be answered ultimately in terms of primrose blossoms, and the universal mind is utterly useless for this purpose unless it can go beyond predicting atomic data and tell also whether they imply or not the future blossoming of the primroses that I have planted today.

I shall not examine for the moment whether we could actually infer something about primroses, or about anything else that we may be interested in, from a knowledge of atomic data. It is enough to have realized that this is not obviously feasible. For it proves that Laplace's ideal of complete scientific knowledge as formulated by him has no immediate bearing on the vast majority of our experiences. Indeed, his representation of the universe ignores as it stands most of our experiences instead of answering our questions in respect to them.

I shall try to show that this essential shortcoming of the Laplacean scheme is irremediable and that it is due to his misunderstanding of the very nature of experimental science.

The point can be introduced by considering the use of a geographical map. A map represents a part of the earth's surface in the same sense in which experimental science represents a much greater variety of experience. As regards the map, it is clear that it functions as such only if used by someone to find his way by it. A map user must be able to do three things. First he must identify his actual position in the landscape with a point on the map, then he must find on the map an itinerary toward his destination, and finally he must identify his itinerary by some landmark in the landscape around him. Three similar stages can be recognized when science is used for the interpretation of experience in the exact predictive sciences of which the Laplacean scheme was an intended idealization. The map is replaced in this case by some formulae, such as for example the laws of planetary motion, which are applied once more in three stages. First we make some measurements which yield a set of numbers, representing our experience at the start; then we operate on these numbers by the aid of our formulae, so as to compute a future event; and finally we look out for the experience predicted by our computation.

The Laplacean ideal of science goes astray in the first place by neglecting or at least trying to render negligible our personal participation in establishing a correspondence between a scientific formalism and the experience to which it is intended to apply. To reveal this mistake we may choose as our example the very province of science—the mechanics of celestial bodies—in which Laplace found his ideal of science. A mathematical theory of planetary motion predicts the position of the planets in numbers, stating their longitude and elevation at any particular time in respect to any particular observatory. From these three numbers we might try to compute the readings on the scales of the observatory's telescope when fixed on the planet at the exact moment when the fingers of a particular clock reach the position corresponding to the time in question. If such a computation were possible it would affirm scientific knowledge without the intervention of any personal act. But it is not possible. We realize this by recalling that any correlation between a measured number introduced into a formula of mechanics and the corresponding instrument readings must rely on an estimate of observational errors, and that this estimate cannot be definitely prescribed by any rule. The indeterminacy in question is due partly to the statistical fluctuations of observational errors but it is even more intractably present in the ever-menacing possibility of systematic errors. For even the most strictly mechanized observational procedure leaves something to personal skill in the exercise of which a personal bias may enter.

We should bear in mind here the famous case of the Astronomer Royal Maskelyne who in the year 1796 dismissed his assistant Kinnebrook for persistently recording the passage of stars more than half a second later than he, his superior. Maskelyne did not realize that an equally watchful observer may register systematically different times by the method employed by him. Not until Bessel revealed this possibility 26 years later was the discrepancy resolved and Kinnebrook belatedly rehabilitated. Despite its reduction through the use of the modern transit micrometer, this type of error is still present in astronomy today. "This *personal equation,*" wrote H. N. Russell (1945), "is an extremely troublesome error, because it varies with the observer's physical condition and also with the nature and brightness of the object. Faint stars are almost always observed too late, in comparison with bright ones; this gives rise to the so-called *Magnitude* equation" (p. 63). We must accordingly allow that some trace of a hidden personal bias may always affect the result of a series of readings. Therefore, even classical mechanics, the branch of science most closely approximating Laplace's ideal, must rely on a measure of personal skill and personal judgment for establishing a valid correspondence between its mathematical formalism and the facts of experience to be represented by it.

This seems a small and pedantic amendment, but it introduces a principle which carries a whole world of implications. It acknowledges that a mathematical formalism can be said to convey knowledge of experience only by presupposing ourselves, the holders of this knowledge, who possess an exceptional delicacy of eye, ear, and touch, developed by personal experience. Thus even an exact science is seen to include an art, the art of establishing correspondence with the raw experience given to our senses.

It is of the essence of an art as the term is used here that it cannot be specified in detail and can therefore not be transmitted by prescription, but only passed on to an apprentice following the practice of a master. Such personal transmission is cumbersome and uncertain as compared with the transmission of measured data. Scientists and technologists are accordingly always striving to depersonalize knowledge as far as possible by representing it in terms of measured quantities, and wherever we see them still rely on connoisseurship, we may assume that it has not been possible to replace it by measurement. The large part of their time spent by students of chemistry, biology, and medicine in acquiring connoisseurship by attending practical courses therefore shows directly how greatly these sciences rely on personal knowledge that is not specifiable.

Moreover, the fundamental concepts of these sciences are drawn from everyday experience in which measurement plays no part. The existence of animals was not discovered by zoologists, nor that of plants by botanists. We learn to distinguish living beings from inanimate matter long before we study biology and we continue to use our original conception of life within biology. Psychologists must know from ordinary experience what intelligence is before they can devise tests for measuring it scientifically. They were ordinary people who, knowing the sufferings of sickness and the joy of recovery, set medical science its task.

It is true that the progress of science is ever moulding and modifying our everyday conceptions. But when this is allowed for it still remains true that there is a vast range of everyday knowledge, based on delicate and complex conceptions, which serves as a guide to biology, medicine, psychology, and to the manifold disciplines which study man and society. This knowledge is transmitted personally to the child by his seniors as he grows up, in the manner of a practical art, in the very same way as a student is taught scientific skills and connoisseurship at the bedside, in the laboratory, or on geological excursions.

This brings out squarely the general principle which limits the scope of the exact sciences of which the Laplacean ideal is an extreme idealization. Most of the questions in which we are interested are of the same kind as that about the blossoming of newly planted primroses. Answers to these questions must be given in terms of *personal knowledge* available to the layman, as corrected and expanded by sciences relying to a considerable extent on the personal knowledge of experts. Laplacean predictions would convey none of this personal knowledge and would therefore ignore almost the entire range of existing knowledge. Hence, if the Laplacean vision, or more generally the exact sciences, insisted on claiming to constitute the whole of knowledge they would impose on us universal ignorance.

It follows that we must revise our ideal of science by acknowledging skills and connoisseurship as valid, indispensable, and definitive forms of knowledge. I shall try to show how this should lead to a far-reaching relaxation of the tension between the exact sciences and other branches of scientific knowledge and may lead on further to a reconciliation between science and the nonscientific aspects and concerns of man.

The principles involved in the arts of knowing are quite similar, whether it is a matter (to use the distinction of Professor Ryle) of "knowing how" or of "knowing what," but they are more easily illustrated by instances of the first kind, which are usually described as the knowing of a skill.

A striking feature of such knowing is the presence of two different kinds of awareness of things that we are skillfully handling. When I use a hammer to drive in a nail, I attend to both, but quite differently. I *watch* the effects of my strokes as I wield the hammer. I do not feel that its handle has struck my palm but that its head has struck the nail. Yet in another sense I am highly alert to the feelings in palm and fingers holding the hammer. They guide my handling of it effectively, and the degree of attention that I give to the nail is given to the same extent in a different way to these feelings. The difference may be stated by saying that these feelings are not watched *in themselves,* but I watch something else by keeping aware of them. I have a *subsidiary awareness* of the feelings in the palm of my hand which is merged into my *focal awareness* of my driving in the nail.

We may think of the hammer replaced by a probe, used for exploring the interior of a hidden cavity. Think how a blind man feels his way by use of a stick, which involves transposing the shocks transmitted to his hand and the muscles holding the stick into an awareness of the things touched by the point of the stick. We have here the transition from knowing *how* to knowing *what* and can see how similar is the structure of the two.

The difference between subsidiary and focal awareness is closely allied to another fundamental distinction, namely that between parts of my own body and things external to it. For we attend to external objects by being subsidiarily aware of things happening within our body. The physiologist finds that the localization of an object in space is based on a slight difference between the two images thrown on our retina, on the accommodation of the eyes, on the convergence of their axis and the effort of muscular contraction controlling the eye motion, supplemented by impulses received from the labyrinth, which vary according to the position of the head in space. Of all these we become aware only in terms of our localization of the objects we are gazing at; and in this sense we may be said to be subsidiarily aware of them.

Our subsidiary awareness of tools and probes can be regarded, therefore, as a condition in which they come to form part of our body. The way we use a hammer or a blind man uses a stick shows in fact that in both cases we shift outward the points at which we make contact with things that we observe as objects outside ourselves. While we rely on a tool or a probe, these things cannot be deliberately handled in themselves or critically examined as external objects. Instead, we pour ourselves into them and assimilate them as part of our own existence, uncritically.

We may generalize the conception of this process so as to include the acceptance and use of intellectual tools such as are offered by an interpretative framework and in particular by the formalism of a science. While we rely on such a formalism it is not an object under examination but a tool of observation. For the time being we have identified ourselves with it, and the exercise of our critical faculties, so long as they are exercised by using this formalism, can only serve to strengthen our uncriticized acceptance of it. The skills involved in carrying out the measurements on which the exact sciences rely are an indispensable extension of the *personal knowing* by which we hold the theories of exact science.

III

The wider field of personal knowledge which we all acquire in everyday life, and which a practical scientific training can extend further, may be fitted in here by noting that our two kinds of awareness apply to the relation between parts and wholes. When we acknowledge a muscular skill we accredit the person who has mastered the skill with the capacity for carrying out numerous part movements with a view to the achievement of a comprehensive result. Similarly, when we acknowledge skillful knowledge, such as that of a medical diagnostician, an art dealer, a taxonomist, or a cotton classer, we accredit such experts with the capacity for appreciating a large number of unspecifiable details in terms of a comprehensive entity jointly constituted by them. In both cases we are aware of a multitude of parts in terms of a whole, and this submerging of the parts in the whole may be described as a subsidiary awareness of the parts within a focal awareness of the whole. The kinship between the process of tool using and that of achieving or perceiving a whole has in fact already been so well established by Gestalt psychology that it may be taken for granted without further argument.

But an objection may be raised at this point. If it is affirmed that a personal contribution of our own is present in every scientific affirmation of ours, whether, e.g., in physics or in psychology, it may appear that this modifies our conception of all sciences equally and therefore leaves the question unanswered whether or not, for example, psychology can be reduced to terms of physics. But we find the answer to this question by realizing how vastly the degree of our personal contribution varies between the different sciences. An exact physical measurement can usually be carried out by a laboratory assistant of normal intelligence without extensive apprenticeship. On the other hand years of experience are required to train an expert medical diagnostician or even a cotton classer or a wine taster. In these skills and connoisseurships there is a much ampler personal contribution than in the knowing and applying of an astronomical formula to successive observed positions of some planet. When we say that the former kinds of knowing are *unspecifiable* we assert that their more richly personal knowledge cannot be exhaustively analyzed in the depersonalized terms offered by the exact sciences.

We shall note conversely also that the extension of the part played by personal knowledge expands continuously from the exact sciences to those sciences which rely more heavily on personal skill and connoisseurship. And this opens up an avenue along which we may find the solution to the basic problem that we set ourselves. A further expansion of our personal participation in the act of knowing may also bridge the gap between the scientific method and the study and conduct of human affairs.[1]

I shall try to show how we may start off at least in the direction indicated by this program. The ideal of exact observation postulates the identity of the observing person before and after the observation. By contrast, if someone gets drunk or falls in love, this is not an observation, for these processes affect the whole person. In an act of personal knowing, the knowing person also modifies himself to some extent by pouring himself into things of which he becomes subsidiarily aware in order to become focally aware of a comprehensive feature comprising the same things.

I have so far described this participation of the knowing person as an extension of his body and this has applied to the know-how of a personal skill and the know-what of an expert diagnosis. But though these two are never quite separable and have a common logical structure, there is a difference in the nature of the processes by which the knowing person commits himself in either case, which makes it easier to analyze them in different terms.

When the emphasis of our knowing lies on how to produce a result, the effort of acquiring or skillfully applying such knowl-

[1]Some bearings of the techniques of connoisseurship on ethical problems have been elaborated by Hare (1952, pp. 111ff.) following Urmson (1950). But Urmson deliberately chooses examples of grading which do not convey the recognition of significant entities but have merely subjective standing (like "*super apples*"). His analysis therefore remains peripheral to the perspective explored in the present argument.

edge may be said to be guided by a purpose. It is in the light of this purpose that certain things may be defined as tools and certain movements may be said to be rationally coordinated. The economical and effective achievement of this purpose sets a standard to our skill. By striving for the fulfillment of this standard we pick up in practice, usually without any focal awareness of doing so, the elements of a successful performance. Thus the striving by which we extend our person in achieving a skill is in the nature of a purpose.

When, on the other hand, the emphasis of our knowing lies in recognizing or understanding a thing, the effort of acquiring or skillfully applying such knowledge may be said to be guided by our attention. A biologist, a doctor, an art dealer, a cloth merchant acquire their expert knowledge in part from textbooks, but these texts are useless to them without the accompanying training of the eye, the ear, and the touch. Only by attentively straining their senses can they acquire the right sense or feel for identifying a certain biological specimen, a case of a certain sickness, a genuine painting by a certain master, a cloth of a distinctive quality. By such training the expert develops an exceptional fastidiousness which enables him to act as valuer for certain objects. This is generally true: every act of personal knowing sets up a standard of excellence. While the athlete and the dancer putting forward their best are acting as critics of their own performances, connoisseurs are the acknowledged critics of certain things. A person is acknowledged as a connoisseur only if he is believed to know whether such things fulfill certain standards that are characteristic of their being good specimens of their own kind. All personal knowing appraises what it knows by a standard that it sets to itself.

Thus the observer's participation in the act of knowing has led us to a point where observation assumes the functions of an appraisal by standards which we regard as impersonal. When we acknowledge that a skillful performance is coherent and ingenious in itself, we appeal to standards of coherence and ingenuity to which we attribute universal significance. The operation of the patent law relies successfully on our capacity for appreciating the presence or absence of a certain degree of ingenuity in a new practical procedure. And again, mathematics can be said to exist as a science only if we credit ourselves with the capacity of appreciating reliably the profundity and ingenuity of certain processes of inference.

When appreciating a healthy plant or animal we again do so by a standard to which we attribute universality. Going further, we appreciate in the same way the coherent behavior of animals and their intelligent performances.

At this stage the process by which the observer identifies himself with the parts of a whole in order to appreciate the qualities of its coherence takes on a new character. Rational behavior and feats of intelligence are observed by the identification of an observer with a person whose rationality or intelligence he appraises. Our capacity for understanding another person's actions by entering into his situation and for judging his actions from his own point of view thus appears to be but an elaboration of the technique of personal knowing which generally comes into play whenever we handle objects skillfully or know them by a subsidiary awareness of some framework in terms of which they are represented or of the parts of which they are composed.

This argument has fulfilled at least part of my program. It has shown that the shortcomings of the Laplacean ideal of science can be remedied only by acknowledging our personal knowing as an integral part of all knowledge accepted by us, and this was then shown to involve an act of uncritical assimilation of certain things which enables us critically to appreciate others; from here we were led to a point where our justification of personal knowing results in accrediting our evaluation of human affairs from the point of view of men as sentient, intelligent, and morally responsible beings.

We may go further. We have seen that our personal knowing operates by an expansion of our person into a subsidiary awareness of particulars that are merged into our focal awareness of a whole, and that this manner of living in parts results in our critical appraisal of their coherence. We may therefore also accredit our living within an historical situation and our acceptance of a certain role in it as legitimate guides to our responsible participation in the problems presented by these situations. Science no longer requires, then, that we study man and society in a detached manner, but instead restores us to an acceptance of our position as members of a human society.

The transition from observation to appraisal has been reached here by following up a sequence of increasingly complex and delicate forms of knowing. I believe that this is the path along which the widest contacts can be established between exact science and other domains of the human mind, but I must indicate here a shorter route leading from observation to appraisal which can be found within the exact sciences themselves. The most striking illustration of this consists in the system of theoretical crystallography.

Here we have a theory which applies to the facts of experience without making any unambiguous predictions about them, offering instead a systematic framework for the appreciation of any possible facts. Crystallography predicts that there are 230 geometrically different types of atomic lattices which should manifest themselves in 32 different classes of crystal symmetry. This statement has an intrinsic geometrical interest without reference to experience. The 230 space groups define indefinitely extended repetitive patterns of points in space, and the 32 classes of symmetry define groups of polyhedra derived by certain rules from these patterns. This purely geometrical theory already presupposes an appreciation of symmetry. It determines all possible types of symmetry and distinguishes between higher and lower grades of symmetry. All this would remain true even if no specimen of any crystal could ever be observed in fact. Yet actually the theory has been validated by a wide area of experience. Indeed, it was the existence of solids possessing a peculiar regularity of shape which first stimulated the geometrical study of such shapes and of the possible atomistic patterns from which these shapes could be derived; and subsequently the geometric theory of crystals has controlled the collection, description, classification, and structural analysis of an immense variety of crystals. This has gone on and established crystallographic theory as part of natural science, without ever exposing it to the hazards of refutation by the facts of experience. For the theory merely offers a classification for all possible solids according to certain types of regularity. It prescribes a process of appraising the regularity of any solid specimen in the light of its standards. Our confidence in such a system of appraisals increases with the number of instances in which it has been found distinctly apposite to experience, and is not in the least weakened by the much larger number of instances about which the system can say practically nothing. Crystallographic theory presents us with an elaborately formalized system of personal knowing, lying quite outside the area which Laplace defined as comprising universal knowledge.

We can now supplement our previous conclusions by this lesson. By a stepwise generalization of the structures of skills and connoisseurship we had achieved a foothold within the realm of human responsibilities supported by scientific practice. We can now derive additional support from science for the kind of thought and judgment that seem appropriate to such a situation. Our moral problems and their attempted solutions, problems of jurisprudence or of art criticism, can no longer be regarded as of questionable significance, on account of the fact that they say nothing about experience that could be true or false. The validity which we ascribe to crystallographic theory demonstrates our capacity for discovering exact standards that are apposite to the appraisal of experience and also our capacity for recognizing their authenticity in this respect.

The line of thought which I have pursued here suggests that we can reconcile science to man by restoring science to its true purpose. I have tried to reveal some of the features of science which, when fully acknowledged, indicate the main steps by which this program may be carried out.

Science does not require that we study man and society in a detached manner. On the contrary, the part played by personal knowledge in science suggests that the science of man should rely on greatly extended uses of personal knowing.

A personal knowledge of man may consist in putting ourselves in the place of the persons we are studying and in trying to solve their problems as they see them or as we see them. That opens the door for our entry into human personality in its whole moral, religious, and artistic outlook, as the bearer of a historical consciousness, a political and legal responsibility. Thus it introduces us through an extension of scientific inquiry straight into the whole sentient, creative, and responsible life of human concerns.

A system of ethics or a code of laws can no longer be regarded as unscientific in a derogatory sense because it predicts nothing that could be true or false, for science is seen to accredit us with the capacity for authentic appreciation of other values than the truth or falsity of a statement. As we know order from disorder, health from sickness, the ingenious from the trivial, we may distinguish with equal authority good from evil, charity from cruelty, justice from injustice.

7

FROM COPERNICUS TO EINSTEIN

In the Ptolemaic system, and in the cosmogony of the Bible, man was assigned a central position in the universe from which he was ousted by Copernicus. Ever since, writers eager to drive the lesson home have urged us, resolutely and repeatedly, to abandon all sentimental egoism, and to see ourselves objectively in the true perspective of time and space. What precisely does this mean? Imagine a full "main feature" film, recapitulating faithfully the history of the universe: the rise of human beings from the first beginnings of man to the achievements of the twentieth century would be covered in it by a single second. Alternatively, if we decided to examine the universe "objectively" in the sense of paying equal attention to portions of equal mass, this would result in a lifelong preoccupation with interstellar dust, relieved only at brief intervals by a survey of incandescent masses of hydrogen–not in a thousand million lifetimes would the turn come for giving man even a second's notice.

It goes without saying that no one–scientists included–looks at the universe this way, whatever lip service is given to "objectivity." Nor should this surprise us. For as human beings, we must inevitably see the universe from a center lying within ourselves and speak about it in terms of a human language shaped by the exigencies of human intercourse; so that any attempt rigorously to eliminate our human evaluation from our picture of the world must lead to absurdity.

First published in *Encounter,* 5(Sept.):1-10, 1955.

What is the true lesson of the Copernican revolution? Why did Copernicus exchange his actual terrestrial station for an imaginary solar standpoint? The only justification for this lay in the greater intellectual satisfaction he derived from the celestial panorama as seen from the sun instead of the earth. Copernicus gave preference to man's delight in abstract theory, at the price of rejecting the evidence of our senses which present us with the irresistible fact of the sun, the moon, and the stars rising daily in the east to travel across the sky toward their setting in the west. In a literal sense, therefore, the new Copernican system was as anthropocentric as the Ptolemaic view, the difference being merely that it preferred to satisfy a different human inclination.

It becomes legitimate to regard the Copernican system as more "objective" than the Ptolemaic only if we accept this very shift in the nature of intellectual satisfaction as the criterion of greater objectivity. This would imply that, of two forms of knowledge, we should consider as more objective that which relies to a greater measure on theory instead of on more immediate sensory evidence. So that, the theory being placed like a screen between our senses and the things of which our senses otherwise would gain a more immediate impression, we would rely increasingly on theoretical guidance for the interpretation of our experience, and would correspondingly reduce the status of our raw impressions to that of dubious or altogether misleading appearances.

It seems to me that we have sound reasons for thus considering theoretical knowledge as more objective than immediate experience. A theory is something other than myself. It may be set out on paper as a system of rules, and it is the more truly a theory the more completely it can be put down in such terms. Mathematical theory reaches the highest perfection in this respect. But even a geographical map fully embodies in itself a set of strict rules for finding one's way through a region of otherwise uncharted experience. Indeed, all theory may be regarded as a kind of map extended over space and time. It seems obvious that a map can be correct or mistaken, so that to the extent to which I have relied on my map I shall attribute to it any mistakes that I made by doing so. A theory on which I rely is therefore objective knowledge in so far as it is not I, but the theory, which is proved right or wrong when I use such knowledge.

I can smell onions, have a headache, or feel humiliated: all these are states of consciousness; but my theory cannot be in any of these states; it exists like a stick or a stone, without any consciousness. Hence a theory cannot be led astray by illusions. To find my way by a map I must perform the conscious act of map reading and may be deluded in the process, but the map cannot be deluded and remains right or wrong in itself, impersonally. Consequently, a theory on which I rely has a rigid formal structure, on the operation of which I can depend whatever mood or desire may possess me.

Since the formal affirmations of a theory are unaffected by the state of the persons accepting it, theories may be constructed without regard to our normal approach to experience. Here is a third reason why the Copernican system, being more theoretical than the Ptolemaic, is also more objective. Since its picture of the solar system disregards our terrestrial location it may claim equal interest to the inhabitants of Mars, Venus, or Neptune, and will likewise commend itself to them–provided that they share our intellectual values.

Thus, when we claim greater objectivity for the Copernican theory we do imply that its excellence is not a matter of personal taste on our part, but an inherent quality of it deserving universal acceptance by rational creatures. We abandon the cruder anthropocentrism of our senses–but only in favor of a more ambitious anthropocentrism of our reason.

Actually, the theory that the planets move round the sun was to speak for itself in a fashion that went far beyond asserting its own inherent rationality. It was to speak to Kepler (66 years after the death of Copernicus) and inspire his discovery of the elliptic path of planets and of their constant angular surface velocity, and to inspire again, 10 years later, his discovery of the third law of planetary motion, relating orbital distances to orbital periods. And, another 68 years later, Newton was to announce to the world that these laws were but an expression of the underlying fact of general gravitation. All these consequences were partially anticipated by Copernicus in the act of discovering the superior rationality of the heliocentric system.

Thus the intellectual satisfaction which the system originally provided, and which gained acceptance for it, proved to be the token of a deeper significance unknown to its originator. Unknown but not entirely unsuspected; for those who wholeheartedly embraced the Copernican system at an early stage (as did Galileo) committed themselves thereby to the expectation of an indefinite range of possible future confirmations of the theory, and this expectation was essential to their belief in the objective validity of the system.

One may say, indeed, quite generally that a theory which we acclaim as rational in itself is thereby accredited with prophetic powers. We accept it in the hope of making contact with reality; so that, being really true, our theory may yet show forth its truth through future centuries in ways undreamed of by its authors. Some of the greatest scientific discoveries of our age have been rightly described as the amazing confirmations of accepted scientific theories. In this wholly indeterminate scope of its true implications lies the deepest sense in which objectivity is attributed to a scientific theory.

Here, then, are the true characteristics of objectivity as exemplified by the Copernican theory. We see that objectivity does not demand that we estimate man's significance in the universe by the minute size of his body, the brevity of his past history, or his probable future career. It does not require that we see ourselves as a mere grain of sand in a million Saharas. It inspires us, on the contrary, with the hope of overcoming the appalling disabilities of our bodily existence, even to the point of conceiving a rational idea of the universe which can authoritatively speak for itself. It is not a counsel of self-effacement but the very reverse–a call to the Pygmalion in the mind of man.

This is not what we are taught today. To say that the discovery of objective truth in science consists in the apprehension of a rationality which commands our respect and arouses our contemplative admiration; that such discovery, while using the experience of our senses as clues, transcends experience by embracing the vision of a reality beyond the impression of our senses, a vision which speaks for itself in guiding us to an ever deeper understanding of reality–such an account of scientific procedure would be shrugged aside as outdated Platonism: mystery mongering, unworthy of an enlightened age.

Yet it is precisely on this conception of objectivity that I wish to insist here. I want to show how scientific theory came to be reduced in the modern mind to the rank of a convenient contrivance, a device for recording events and computing their future course, and then to suggest that twentieth-century physics, and Einstein's discovery of relativity in particular, which are usually regarded as the fruits and illustrations of this positivistic conception of science, demonstrate on the contrary the power of science to make contact with reality in nature by recognizing what is rational in it.

To show this, I have first to go back to a previous stage of science which begins long before Copernicus, though it leads straight up to him. It starts with Pythagoras, who lived a century before Socrates. Even so, Pythagoras was a latecomer in science, for the scientific movement was started almost a generation earlier, on rather different lines, by the Ionian school of Thales. Pythagoras and his followers did not, like the Ionians, try to describe the universe in terms of certain material elements, but interpreted it exclusively in terms of numbers. They took numbers to be the ultimate substance, as well as the form, of things and processes. When sounding an octave they could hear the simple numerical ratio of 1:2 in the harmonious chiming of the sounds from two wires whose lengths had the ratio 1:2. Thus, acoustics made the perfection of simple numerical relations audible to their ear. They turned their eyes toward the heavens and saw the perfect circle of the sun and moon; they watched the diurnal rotation of the firmament and, studying the planets, saw them governed by a complex system of steady circular motions; and they apprehended these celestial perfections in the way one listens to a pure musical interval. They listened to the music of the spheres in a state of mystic communion.

The revival of astronomical theory by Copernicus after two millennia was a conscious return to the Pythagorean tradition. While studying law in Bologna he worked with the professor of astronomy, Novara, a leading Platonist who taught that the universe was to be conceived in terms of simple mathematical relationships.

Then, on his return to Cracow, with the thought of a heliocentric system in his mind, he made a further study of the philosophers and traced his new conception of the universe back to writers of antiquity standing in the Pythagorean tradition. He, and after him Kepler, continued wholeheartedly the Pythagorean quest for harmonious numbers and geometrical excellence.

In the volume containing the first statement of his third law, we can see Kepler speculating intensely on the way the sun, which is the center of the cosmos and therefore somehow *nous* (reason) itself, apprehends the celestial music performed by the planets: "Of what sort vision is in the sun, what are its eyes, or what other impulse it has...even without eyes...for judging the harmonies of the [celestial] motions," it would be "for those inhabiting the earth, not easy to conjecture"–yet one may at least dream, "lulled by the changing harmony of the band of planets," that "in the sun there dwells an intellect simple, intellectual fire or mind, whatever it may be, the fountain of all harmony." He went so far as to write down the tune of each planet in musical notation.

To Kepler astronomic discovery was ecstatic communion, as he voiced it in a famous passage of the same work:

> What I prophesied two-and-twenty years ago, as soon as I discovered the five solids among the heavenly orbits–what I firmly believed long before I had seen Ptolemy's Harmonics–what I had promised my friends in the title of this fifth book, which I named before I was sure of my discovery–what sixteen years ago I urged to be sought–that for which I have devoted the best part of my life to astronomical contemplations, for which I joined Tycho Brahe...at last I have brought it to light, and recognized its truth beyond all my hopes....So now since eighteen months ago the dawn, three months ago the proper light of day, and indeed a very few days ago the pure Sun itself of the most marvellous contemplation has shone forth–nothing holds me; I will indulge my sacred fury; I will taunt mankind with the candid confession that I have stolen the golden vases of the Egyptians, in order to build of them a tabernacle to my God, far indeed from the bounds of Egypt. If you forgive me, I shall rejoice; if you are angry, I shall bear it; the die is cast, the book is written, whether to be read now or by posterity I care not: it may wait a hundred

years for its reader, if God himself has waited six thousand years for a man to contemplate His work.

What Kepler claimed here about the Platonic bodies was nonsense, and his exclamation about God's having waited for him for thousands of years sounds fanciful; yet his outburst conveys a true idea of the scientific method and of the nature of science; an idea which has since been disfigured and lost by the sustained attempt to remodel it in the likeness of a mistaken ideal of objectivity.

Passing from Kepler to Galileo we see the transition to a dynamics in which for the first time numbers enter as measured quantities into mathematical formulae. But with Galileo this usage still applies only to terrestrial events, while in respect to heavenly motions he still holds the Pythagorean view that the book of nature is written in geometrical characters. In *The Two Great Systems of the World*, he argues in the Pythagorean tradition from the principle that the parts of the world are perfectly ordered. He still believes that the motion of the heavenly bodies—in fact all natural motion as such—must be circular. Rectilinear motion implies change of place, and this can occur only from disorder to order: that is, either in the transition from primeval chaos to the right disposition of the parts of the world, or in violent motion, i.e., in the endeavor of a body artificially moved to return to its "natural" place. Once the world order is established, all bodies are "naturally" at rest or in circular motion. Uniform motion along a straight line, on which Galileo himself founded modern dynamics, appears only in the interstices of a Pythagorean universe.

Thus the first century after the death of Copernicus was inspired by Pythagorean intimations. Their last great manifestation was perhaps Descartes's universal mathematics: his hope of establishing scientific theories by the apprehension of clear and distinct ideas, which as such were necessarily true.

But a different line of approach was already advancing gradually, stemming from the other line of Greek thought, which lacked the mysticism of Pythagoras, and which recorded observations of all kinds of things, however imperfect. This school derived from the Ionian philosophers culminated in Democritus, a contemporary of Socrates, who first taught men to think in materialistic terms. He laid down the principle: "By convention sweet, by convention bitter, by convention cold, by convention hot, by convention colored; in reality only atoms and the void." With this Galileo himself agreed; the mechanical properties of things alone were primary (in Locke's terminology), their other properties were derivative, or secondary. Eventually it would appear that the primary qualities of such a universe could be brought under intellectual control by applying Newtonian mechanics to the motions of matter, while its secondary qualities could be derived from this underlying primary reality. Thus emerged the mechanistic conception of the world which prevailed virtually unchanged till the end of the last century.

This was also a theoretical and objective view in the sense of replacing the evidence of our senses by a formal space-time map that predicted the motions of the material particles which were supposed to underly all sensual experience. Yet there is a definite change from the Pythagorean to the Ionian conception of theoretical knowledge. Theory no longer reveals perfection; it no longer contemplates the harmonies of Creation. Numbers and geometrical forms are no longer assumed to be inherent as such in nature. In Newtonian mechanics the formulae governing the mechanical substratum of the universe were differential equations containing no numerical rules and exhibiting no geometrical symmetry. Henceforth "pure" mathematics became strictly separated from the *application* of mathematics to the formulation of empirical laws. Geometry became the science of empty space; and analysis, affiliated since Descartes to geometry, seceded with it into the region beyond experience. Mathematics represented all rational knowledge which appeared necessarily true; while the events of the world were seen as contingent–that is, merely such as happened to be the case.

The separation of reason and experience was pressed further by the discovery of non-Euclidian geometry. Mathematics was thereafter denied the capacity of stating anything beyond sets of tautologies formulated within a conventional framework of notations. A new positivist philosophy arose, denying to the scientific theories of physics any claim to inherent rationality, a claim which it condemned as metaphysical and mystical. The earliest, most energetic and influential development of this idea was due to Ernst

Mach, who, by his book *Die Mechanik,* published in 1883, founded the Viennese school of positivism. Scientific theory, according to Mach, is merely a convenient summary of experience. Its purpose is to save time and trouble in recording observations. It is the most economical adaptation of thought to facts, and just as external to the facts as a map, a timetable, or a telephone directory; indeed this conception of scientific theory would include a timetable or a telephone directory among scientific theories.

Accordingly, scientific theory was denied all persuasive power that is intrinsic to itself. It must not go beyond experience by affirming anything that cannot be tested by experience; and above all, scientists must be prepared immediately to drop a theory the moment an observation turns up which conflicts with it. In so far as a theory cannot be tested by experience–or appears not capable of being so tested–it ought to be revised so that its predictions are restricted to observable magnitudes.

This view, which can be traced back to Locke and Hume and which in its massive modern absurdity has almost entirely dominated twentieth-century thinking on science, seems to be the inevitable consequence of separating, in principle, mathematical knowledge from empirical knowledge. I shall now proceed to the story of relativity, which is supposed to have brilliantly confirmed this view of science, and shall show why in my opinion it has, on the contrary, supplied some striking evidence for its refutation.

The story of relativity is a complicated one, owing to the currency of a number of historical fictions. The first of these can be found in every textbook of physics. It tells you that relativity was conceived by Einstein in 1905 in order to account for the negative result of the Michelson-Morley experiment, carried out in Cleveland 18 years earlier, in 1887. Michelson and Morley are alleged to have found that the speed of light measured by a terrestrial observer was the same in whatever direction the signal was sent out. That was surprising, for one would have expected that the observer would to some extent catch up with signals sent out in the direction in which the earth was moving, so that the speed would appear slower in this direction, while the observer would move away from the signal sent out in the opposite direction so that the speed would then appear faster.

The experiment is supposed to have shown no trace of such an effect due to terrestrial motion, and so—the textbook story goes on—Einstein undertook to account for this by a new conception of space and time, according to which we could expect invariably to observe the same value for the speed of light, whether we are at rest or in motion. So Newtonian space, which is "necessarily at rest without reference to any external object," and the corresponding distinction between bodies in absolute motion and bodies at absolute rest, were abandoned and a framework set up in which only the relative motion of bodies could be expressed.

But the historical facts are different. Einstein had already speculated as a schoolboy, at the age of 16, on the curious consequences that would occur if an observer pursued and kept pace with a light signal sent out by him. His autobiography reveals that he discovered relativity

> ...after ten years reflection...from a paradox upon which I had already hit at the age of sixteen: If I pursue a beam of light with the velocity c (velocity of light in a vacuum), I should observe such a beam of light as a spatially oscillatory electromagnetic field at rest. However, there seems to be no such thing, whether on the basis of experience or according to Maxwell's equations. From the very beginning it appeared to me intuitively clear that, judged from the standpoint of such an observer, everything would have to happen according to the same laws as for an observer who, relative to the earth, was at rest [p. 53].

There is no mention here of the Michelson-Morley experiment; its findings were, on the basis of pure speculation, rationally intuited by Einstein before he had ever heard about it. To make sure of this, I addressed an inquiry to the late Professor Einstein, who confirmed that "the Michelson-Morley experiment had a negligible effect on the discovery of relativity."

Actually, Einstein's original paper announcing the special theory of relativity (1905) gave little grounds for the current misconception concerning the origins of his discovery. It opens with a long paragraph referring to the anomalies in the electrodynamics of moving media, mentioning in particular the lack of symmetry in its treatment, on the one hand, of a wire with current flowing through it moving relative to a magnet at rest, and on the other of a magnet moving relative to the same electric current at rest. It then goes on to say that "similar examples, as well as the unsuccessful attempts to observe the relative motion of the earth in respect to the medium of light lead to the conjecture that, as in mechanics, so also in electrodynamics, absolute rest is not observable."

The usual textbook account of relativity as a theoretical response to the Michelson-Morley experiment is an invention. It is the product of prejudice, exactly on a par, for example, with the notion customary among primitive people, that hostile witchcraft may be assumed to account for someone's violent death. Even as the native refuses to accept the possibility of accidental death, the modern positivist refuses to acknowledge man's inherent power to discover rationality in nature, and when his prejudices come in conflict with experience, the positivist–like the savage–automatically supplements experience from the resources of his imagination. So when Einstein discovered rationality in nature, unaided by any observation that had not been available for at least 50 years before, our textbooks promptly covered up the scandal by an appropriately embellished account of his discovery.

There is an aspect of this story that is even more curious. For the program which Einstein carried out was largely prefigured by the very positivist conception of science which his own achievement so flagrantly refuted. It was formulated explicitly by Ernst Mach, who, as we have seen, had first advanced the conception of science as a timetable or telephone directory. He had extensively criticized Newton's definition of space and absolute rest on the grounds that it said nothing that could be tested by experience. He condemned this as dogmatic, since it went beyond experience, and as *meaningless,* since it pointed to nothing that could conceivably be tested by experience. Mach urged that Newtonian dynamics should be reformulated so as to avoid referring to any movement of bodies except as the relative motion of bodies with respect to each other, and Einstein acknowledged the "profound influence" which Mach's book exercised on him as a boy and subsequently on his discovery of relativity.

Yet if Mach had been right in saying that Newton's conception of space as absolute rest was meaningless–because it said nothing that could be proven true or false—then Einstein's rejection of Newtonian space could have made no difference to what we hold to be true or false. It could not have led to the discovery of any new facts. Actually, Mach was quite wrong: he forgot about the propagation of light and did not realize that in this connection Newton's conception of space was far from untestable. Einstein, who realized this, showed that the Newtonian conception of space was not *meaningless* but *false.*

Mach's great merit lay in possessing an intimation of the mechanical universe in which Newton's assumption of a single point at absolute rest was eliminated. His was a super-Copernican vision, totally at variance with our habitual experience. Remember that every object we perceive is set off by us instinctively against a background which is taken to be at rest. To set aside this urge of our senses, which Newton had embodied in his axiom of an "absolute space" said to be "inscrutable and immovable," was a tremendous step toward a theory grounded in reason and transcending the senses. Its power lay precisely in that appeal to rationality which Mach wished to eliminate from the foundations of science. No wonder therefore that he advanced it on false grounds, attacking Newton for making an empty statement and overlooking the fact that far from being empty, the statement was false. Thus Mach prefigured the great theoretic vision of Einstein, sensing its inherent rationality, even while trying to exorcize the very capacity of the human mind by which he gained this insight.

But there yet remains an almost ludicrous part of the story to be told. The Michelson-Morley experiment of 1887, which Einstein mentions in support of his theory and which the textbooks have since enshrined as the crucial evidence which compelled him to formulate it, actually did not give the result required by relativity! It admittedly substantiated its authors' claim that the relative motion of the earth and the "ether" did not exceed a quarter of the earth's orbital velocity. But the actually observed effect was not negligible; or has, at any rate, not been proved negligible up to this day. The presence of a positive effect in the observations of Michelson and Morely was pointed out first by W. M. Hicks in 1902 and was later evaluated by D. C. Miller as corresponding to an "ether drift" of eight to nine kilometers per second. Moreover, an effect of the same magnitude was reproduced by D. C. Miller and his collaborators in a long series of experiments extending from 1902 to 1926, in which they repeated the Michelson-Morley experiment with new, more accurate apparatus, many thousands of times.

The layman, taught to revere scientists for their absolute respect for the observed facts, and for the judiciously detached and purely provisional manner in which they hold scientific theories (always ready to abandon a theory at the sight of any contradictory evidence), might well have thought that, at Miller's announcement of this overwhelming evidence of "a positive effect" in his presidential address to the American Physical Society on December 29th, 1925, his audience would have instantly abandoned the theory of relativity. Or, at the very least, that scientists–wont to look down from the pinnacle of their intellectual humility upon the rest of dogmatic mankind–might suspend judgment in this matter until Miller's results could be accounted for without impairing the theory of relativity. But no; by this time they had so well closed their minds to any suggestion which threatened the new rationality achieved by Einstein's world picture that it was almost impossible for them to think again in different terms. Little attention was paid to the experiments; the evidence was set aside, in the hope that it would one day turn out to be wrong.[1]

The experience of D. C. Miller demonstrates quite plainly the hollowness of the assertion that science is simply based on experiments which anybody can repeat at will. It shows that any critical verification of a scientific statement requires the same powers for recognizing rationality in nature as does the process of scientific discovery, even though it exercises these at a lower level. When philosophers analyze the verification of scientific laws, they invariably choose as specimens those which are not in doubt, and thus inevitably overlook the intervention of these powers; for they are describing only the practical demonstration of a scientific law, and

[1]It is true that in the subsequent years experiments by Kennedy, Illingworth, and Michelson, Pease, and Pearson, conducted by different methods, gave essentially null results, but the original Michelson-Morley experiment of 1887 and its innumerable repetitions by D. C. Miller remained and still remain unaccounted for. While desultory attempts to resolve this anomaly continue to be made up to this day, it has never affected the standard textbook accounts of the Michelson-Morley experiment or disturbed in the least the continued teaching of the story of its decisive effect on the discovery of relativity. Still less has it discouraged so far the philosophy of science fostered by this myth; a myth which is itself an emanation of this philosophy.

not its critical verification. As a result we are given an account of scientific method which, having left out the process of discovery on the grounds that it follows no definite method, overlooks the process of verification by referring only to examples where no real verification takes place.

At the time that Miller announced his results, relativity had yet made few predictions that could be confirmed by experiment. Its empirical support lay mainly in a number of already known observations. The account which the new theory gave of these known phenomena was considered as rational, since it derived them from one single convincingly rational principle. It was the same as when Newton's comprehensive account of Kepler's three laws, of the moon's period, and of terrestrial gravitation–in terms of a general theory of universal gravitation–was immediately given a position of surpassing authority, even before any predictions had been deduced from it. It was this inherent rational excellence of relativity which moved Max Born, despite the strong empirical emphasis of his account of science, to salute as early as 1920 "the grandeur, the boldness, and the directness of the thought" of relativity, which made the world picture of science "more beautiful and grander."

Since then the passing years have brought wide and precise confirmation of at least one formula of relativity; probably the only formula ever sent sprawling across the cover of *Time* magazine. The reduction of mass *(m)* by the loss of energy *(e)* accompanying nuclear transformation has been repeatedly shown to confirm the now famous equation $e = mc^2$, where c is the velocity of light. But such verifications of relativity are but confirmations of the earlier judgment of Einstein and his followers, who committed themselves to the theory long before these verifications. And they are an even more remarkable justification of the earlier strivings of Ernst Mach for a more rational foundation of mechanics, setting out a program for relativity at a time when no avenues could yet be seen toward this objective.

The beauty and power inherent in the rationality of contemporary physics is, as I have said, of a novel kind. When classical physics first superseded the Pythagorean tradition, mathematical theory was reduced to an instrument for computing the mechanical motions which were supposed to underlie all natural phenomena. Geometry also stood outside nature, claiming to offer an a priori analysis of Euclidean space, which was regarded as the scene of all natural phenomena but not thought to be involved in them. Relativity, and subsequently quantum mechanics and modern physics generally, have moved back toward a mathematical conception of reality. Essential features of the theory of relativity were anticipated as mathematical problems by Riemann in his development of non-Euclidean geometry; while its further elaboration relied on the powers of the hitherto purely speculative tensor calculus, which by a fortunate accident Einstein got to know from a mathematician in Zurich. Similarly, Max Born happened to find the matrix calculus ready to hand for the development of Heisenberg's quantum mechanics, which could otherwise never have reached concrete conclusions. These examples could be multiplied. By them, modern physics has demonstrated the power of the human mind to discover and exhibit a rationality which governs nature, before ever approaching the field of experience in which previously discovered mathematical harmonies were to be revealed as empirical facts.

Thus relativity has restored, up to a point, the blend of geometry and physics which Pythagorean thought had first naïvely taken for granted. We now realize that Euclidean geometry, which until the advent of general relativity was taken to represent experience correctly, referred only to comparatively superficial aspects of physical reality. It gave an idealization of the metric relations in rigid bodies and elaborated these exhaustively, while ignoring entirely the masses of the bodies and the forces acting on them. The opportunity to expand geometry so as to include the laws of dynamics was offered by its generalization into many-dimensional and non-Euclidean space and this was accomplished by work in pure mathematics, before any empirical investigation could even be imagined. The first step was taken by Minkowski in 1908 by presenting a geometry which expressed the special theory of relativity, and which included classical dynamics as a limiting case. The laws of physical dynamics now appeared as geometrical theorems of a four-dimensional non-Euclidean space. Subsequent investigations by Einstein led, by a further generalization of this type of geometry, to the general theory of relativity, its postulates being so chosen as to produce invariant expressions with regard to all frames of reference assumed to be physically equivalent. As a result of these postulates, the trajectories of masses follow geodetics, and light is propagated along zero lines. When the laws of physics thus appear as particular instances of geometrical theorems, we may infer that the confidence placed in physical theory owes much to its possessing the same kind of excellence from which pure geometry and pure mathematics in general derive their interest, and for the sake of which they are cultivated.

We cannot truly account for our acceptance of such theories without endorsing our acknowledgment of a beauty that exhilarates and a profundity that entrances us. Yet the prevailing conception of science, based on the disjunction of subjectivity and objectivity, seeks—and must seek at all costs—to eliminate such emotional appraisals of theories from science, or at least to minimize their function to that of a negligible byplay. For modern man has set up as the ideal of knowledge the conception of natural science as a set of statements which are "objective" in the sense that their substance is entirely determined by observation, even while their presentation may be shaped by convention. This conception, stemming from a craving rooted in the very core of our culture, would be shattered if the assessment of rationality in nature had to be acknowledged as a justifiable and indeed quite essential part of scientific theory. That is why scientific theory is represented as a mere economical description of facts; or as embodying a conventional policy for drawing empirical inferences; or as a working hypothesis, suited to man's practical convenience: interpretations which all deliberately overlook the rational core of science.

That is why also, if the existence of this rational core yet reasserts itself, its offensiveness is covered up by a set of euphemisms, a kind of decent understatement like that used in Victorian times when legs were called limbs; a bowdlerization which we may observe, for example, in the attempts to replace "rationality" by "simplicity." It is legitimate, of course, to regard simplicity *as a mark* of rationality, and to pay tribute to any theory as a triumph of simplicity. But great theories are rarely simple in the ordinary sense of the term—both quantum mechanics and relativity are very difficult to understand: it takes only a few minutes to memorize the facts accounted for by relativity, but years of study may not suffice to master the theory and see these facts in its context.

Hermann Weyl (1949) lets the cat out of the bag by saying: "... the required simplicity is not necessarily the obvious one but we must let nature train us to recognize the true inner simplicity" (p. 155). In other words, simplicity in science can be made equivalent to rationality only if "simplicity" is used in a special sense known solely by scientists. We understand the meaning of the term "simple" only by recalling the meaning of the term "rational" or "reasonable" or "such that we ought to assent to it," which the term "simple" is supposed to replace. The term "simplicity" functions, then, merely as a disguise for another meaning than its own. It is used for smuggling an essential quality into our appreciation of a scientific theory, which a mistaken conception of objectivity forbids us openly to acknowledge.

What has just been said of "simplicity" applies equally to "symmetry" and "economy." They are contributing elements in the excellence of a theory, but can account for its merit only if the meaning of these terms is stretched far beyond their usual scope so as to include by way of pseudo substitution the much deeper qualities which make the scientists rejoice in a vision like that of relativity; they must stand for those peculiar intellectual harmonies, which reveal more profoundly and permanently than any sense experience, the presence of objective truth.

I could go on like this and criticize pragmatism and operationalism in similar terms. For they are but further variants of the same unending effort to eliminate the essential contribution by which the scientist emotionally participates in his knowledge of a scientific theory. The method of these attempts is always essentially the same: namely, to define scientific merit in trivial terms which are then made to function in precisely the same way as the true terms which they are supposed to eliminate. This is what I have called pseudo substitution; a disingenuous game of playing down man's real and indispensable intellectual powers for the sake of maintaining a set of sterile prejudices.

There are other sections of science which illustrate even more effectively the part played by what might be called personal knowledge in our understanding of nature. Inexact sciences rely heavily on skills and connoisseurship; so also does the appreciation of probability and order in the exact sciences. At all these points science relies on human *appraisal.* The personal appraisal which enters into knowledge denies in a sense the disjunction between subjectivity and objectivity, for it claims that man can transcend his own subjectivity by passionately striving to fulfill his own personal obligations to universal standards. Our endorsement of personal knowledge re-establishes man's responsibility for scientific knowledge on the grounds that our passionate participation in the act of knowing is intrinsic to it, and that it can yet fulfill universal demands.

8. FAITH AND REASON

Recently there fell into my hands—by the kindness of its author—a book which has revealed to me a new, and I think much better, understanding of the situation we are facing today in consequence of the modern scientific revolution. The author's name is Josef Pieper, professor of philosophical anthropology at the University of Münster, and his book which so impressed me is entitled *Scholasticism* (1960).

Owing to this book, I can see now that the conflict between faith and reason evoked by natural science today is but a modern variant of a problem which has filled the thoughts of men in other forms ever since the dawn of philosophic speculation 2,500 years ago.

You will notice that by dating the beginning of philosophy in the sixth century B.C., I am localizing this event in Greece, and more particularly in Ionia and the Greek isles. I know this may be challenged and shall not argue it. Suffice it to say that in my view our anxiety about the relation between faith and reason here in Europe today is the legacy of a particular intellectual family. Modern science has recently been spreading this disturbance all over the planet, but it has formed no part of the heritage of Chinese or Hindu thought. It originated with us here in Europe, and for two and a half millennia it has remained the preoccupation of that part of humanity that has culturally centered on Europe.

But even within these limits, the perspective I now see appearing before me is widely sweeping. I see extending behind us three

First published in *Journal of Religion,* 41:237-247, 1961.

consecutive periods of rationalism, the Greek, the medieval, and the modern. Greek rationalism rose from a bed of mythopoetic thought. Myths and ritual couch most thoughts of men in terms of I-Thou and leave little of importance to be said in terms of I-It. Greek rationalism tended to liberate the mind from this pervasively personal network and to establish in its place broad areas of objective thought. It extended I-It relations into a philosophic interpretation of things.

The Christian message exploded into this scene as an outrage to rationalism. It restored the I-Thou relation to the very center of everything. It proclaimed that a man put to death a few years before in a remote provincial capital was the Son of Almighty God ruling the universe, and he, this man, had atoned by his death for the sins of mankind. It taught that it was the Christian's duty to believe in this epochal event and to be totally absorbed by its implications. Faith, faith that mocks reason, faith that scornfully declares itself to be mere foolishness in the face of Greek rationalism—this is what Paul enjoined on his audiences.

The picture is well known. But you may ask where I see any trace here of a new Christian, medieval rationalism striving to reconcile faith with reason. It emerged later, as the Christian message spread among an intelligentsia steeped in Greek philosophy. It was formulated by Augustine in terms that became statutory for a thousand years after. Reason was declared by him ancillary to faith, supporting it up to the point where revelation took over, after which in its turn faith opened up new paths to reason. What Professor Pieper has shown me for the first time is that the entire movement of scholastic philosophy from Boethius to William of Occam was but a variation on this theme.

Occam brought scholasticism to a close by declaring that faith and reason were incompatible and should be kept strictly separate. Thus he ushered in the period of modern rationalism, which also accepts this separation, but with the new proviso that reason alone can establish true knowledge. Henceforth, as John Locke was soon to put it, faith was no longer to be respected as a source of higher light, revealing knowledge that lies beyond the range of observation and reason, but was to be regarded merely as a personal acceptance which falls short of rational demonstrability. The mutual position of the two Augustinian levels of truth was inverted.

In a way, this step would have brought us back to Greek rationalism, and many of its authors did so regard it. They hoped that the new secular world view would appease religious strife and bring back the blessings of an antique, dispassionate religious indifference. However, post-Christian rationalism soon entered on paths man had never trodden before, and we stand here today at the dismal end of this journey.

But I have not come here to denounce modern rationalism. The arts, the intellectual splendors, and moral attainments of the last 300 years stand unrivaled in the history of mankind. The very failures and disasters that surround us may themselves bear testimony to this. Only gigantic endeavor could precipitate us into such absurdities as the modern scientific outlook has made current today and could set millions ablaze with the peculiar skeptical fanaticism of our age.

Keeping these awful aspects of our situation tacitly in mind, I shall try to trace a new line of thought along which, I believe, we may recover some of the ground rashly abandoned by the modern scientific outlook. I believe indeed that this kind of effort, if pursued systematically, may eventually restore the balance between belief and reason on lines essentially similar to those marked out by Augustine at the dawn of Christian rationalism.

I shall start off in this direction by surveying some essential features of the process of knowing which are disregarded by the modern conception of positive, scientific knowledge.

A few years ago a distinguished psychiatrist demonstrated to his students a patient who was having a mild fit of some kind. Later the class discussed the question whether this had been an epileptic or a hysteroepileptic seizure. The matter was finally decided by the psychiatrist: "Gentlemen," he said, "you have seen a true epileptic seizure. I cannot tell you how to recognize it; you will learn this by more extensive experience."

The psychiatrist knew how to recognize this disease, but he was not at all certain how he did this. In other words, he recognized the disease by attending to its total appearance and did so by relying on a multitude of clues which he could not clearly specify. Thus his knowledge of the disease differed altogether from his knowledge of these clues. He recognized the disease by attending to it, while he was not attending to its symptoms in themselves, but only as clues. We may say that he was knowing the clues only by relying on them for attending to the pathological physiognomy to which they contributed. So if he could not tell what these clues were, while he could tell what the disease was, this was due to the fact that, while we can always identify a thing we are attending to, and indeed our very attending identifies it, we cannot always identify the particulars on which we rely in our attending to the thing.

And this fact can be generalized widely. There are vast domains of knowledge–of which I shall speak in a moment–that exemplify in various ways that we are in general unable to tell what particulars we are aware of when attending to a whole, that is, to a coherent entity which they constitute. Thus we discover that there are two kinds of knowing which invariably enter jointly into any act of knowing a comprehensive entity. There is (1) the knowing of a thing by attending to it, in the way we attend to an entity as a whole, and (2) the knowing of a thing by relying on our awareness of it, in the way we rely on our awareness of the particulars forming the entity for attending to it as a whole.

These two kinds of knowing are not only distinct but also, to an important extent, mutually exclusive. We cannot attend to a clue as a thing in itself without depriving it of its meaning as a clue and thereby losing sight of the thing to which it served as a clue. Gestalt psychology has proved quite generally that we cannot focus our attention on the particulars of a whole without impairing our grasp of the whole; and that, conversely, we can focus on a whole only by reducing our awareness of the particulars to the contribution they make to the whole. We may call the latter a subsidiary awareness of the particulars in terms of our knowledge of the whole that is subserved by them.

As a rule the two alternative kinds of knowing do not completely extinguish each other. We may successfully analyze the symptoms of a disease and concentrate our attention on its several particulars, and then we may return to our conception of its general appearance by becoming once more subsidiarily aware of these particulars as contributing to the total picture of the disease. Indeed, such an oscillation of detailing and integrating is the royal road to deepening our understanding of any comprehensive entity.

In saying this I have pronounced a key word. I have spoken of understanding. Understanding, comprehension–this is the cognitive faculty cast aside by a positivistic theory of knowledge, which refuses to acknowledge the existence of comprehensive entities as distinct from their particulars; and this is the faculty which I recognize as the central act of knowing. For comprehension can never be absent from any process of knowing and is indeed the ultimate sanction of any such act. What is not understood cannot be said to be known.

Let me rapidly run through various forms of knowing to which this analysis can be seen to apply. I have so far used as my leading example the process of medical *diagnostics.* We have a closely similar process in the identification of the species to which an animal or a plant belongs. An expert who can identify 800,000 species of insects must rely on a vast number of clues which he cannot identify in themselves. This is why zoology and botany cannot be learned from printed pages, any more than medicine can. This is why so many hours of practical teaching in the laboratory have to be given in many other branches of the natural sciences also. Wherever this happens, some knowledge of the comprehensive aspect of things is being transmitted: a kind of knowledge which we must acquire by becoming aware of a multitude of clues that cannot be exhaustively identified.

But we hardly ever do such diagnosing without examining the object in question, and this *testing* has itself to be learned along with the art of recognizing the physiognomies of the tested objects. We must jointly learn to be skillful testers as well as expert knowers. Actually, these are only two different and inseparable processes of comprehension. Expert knowing relies on a comprehension of clues, while skillful examination relies on a combination of dexterous motions for tracing these clues.

This reveals the structure of *skills* quite generally. A performance is called skillful precisely because we cannot clearly identify its component muscular acts. The craftsman's cunning consists in controlling these component acts jointly with a view to a comprehensive achievement. Such also is the sportsman's and the musical performer's art. Neither can tell much—and mostly can tell very little—about the several muscular acts he combines in accomplishing his art.

Skills usually require *tools*—instruments of some kind—and these are things akin to the particulars of a comprehensive entity. For they are tools or instruments by virtue of the fact that we rely on them for accomplishing something to which we are attending by using the tool or instrument. In this case we can admittedly identify that on which we rely, though mostly we do not quite know how we actually use it. In any case, it still remains strikingly true that we cannot direct our attention to an object as mere object while relying on it as the tool of a skillful performance. You must keep your eye on the ball, and if you look at your bat instead, you inevitably lose the stroke. Any skillful performance is paralyzed by focal attendance to its particulars, whether these are the dexterous movements of our bodies or the tools which we employ.

The same is true of *speech.* Listen intently to the sound of your own words, disregarding their meaningful context which is the comprehensive entity that they should subserve, and you will be instantly struck dumb. The same is true of the whole multitude of signs, symbols, and gestures by which human communications are achieved and by the practical use of which the intelligence of man is developed far beyond that of the animals. Here is another vital area of skillful doing and knowing, all over which we are met with comprehensive entities to which we can attend only by relying subsidiarily on things and acts of our own to which for the time being we do not attend–and must not attend–in themselves.

Last, deep down, in the most primitive forms of knowing, in the act of *sensory perception,* we meet with the very paradigm of the structure which I have postulated for all kinds of knowledge at all levels. It is sensory perception, and particularly the way we see things, that has supplied Gestalt psychologists with material for their fundamental discoveries which I am expanding here into a new theory of knowledge. They have shown that our seeing is an act of comprehension for which we rely, in a most subtle manner, on clues from all over the field of vision as well as on clues inside our bodies, in the muscles controlling the motion of the eyes and in those controlling the posture of the body. All these clues become effective only if we keep concentrating our attention on the objects we are perceiving. Many of the clues of perception cannot be known in themselves at all; others can be traced only by acute scientific analysis; but all of them can serve the purpose of seeing what is in front of us only if we make no attempt to look at them or attend to them in themselves. They must be left to abide in the role of unspecifiable particulars of the spectacle perceived by our eyes if we are to see anything at all.

This concludes my list. We have now before us the art of *diagnostics* and of the *testing* of objects to be diagnosed, as taught in universities; we have the practice of *skills* in general and the skillful use of *tools* in particular which leads on to the use of *words* and other *signs* by which human intelligence is developed; and finally we have the act of *perception,* the most fundamental manifestation of intelligence, in both animals and men. In each of these cases we have recognized the typical elements of comprehension. I now want to show how this panorama of knowing suggests a new conception of knowledge, equally comprising both the I-It and the I-Thou and at the same time establishing a new harmony between belief and reason.

Clearly, the new element I have introduced here into the conception of knowing is the knowing of things by relying on our awareness of them for attending to something else that comprises them. And we may remember now that there is one outstanding and obvious experience of certain things which we know almost exclusively by relying on them. Our *body* is a collection of such things; we hardly ever observe our own body as we observe an external object, but we continuously rely on it as a tool for observing objects outside and for manipulating these for purposes of our own. Hence we may regard the knowing of something by attending to something else as the kind of knowledge we have of our own body by living in it. This kind of knowing is not an I-It relation but rather a way of existing, a manner of being. We might perhaps call it an I-Myself or I-Me relation.

We are born to live in our body and to feel that we are relying on it for our existence, but the more skillful uses of our body—however elementary—have to be acquired by a process of learning. For example, the faculty of seeing things by using our eyes is not inborn; it has to be acquired by a process of learning.

Hence when we get to know something as a clue, as a particular of a whole, as a tool, as a word, or as an element contributing to perception, by learning to rely on it, we do so in the same way as we learn to rely on our body for exercising intellectual and practical control over objects of our surroundings. So any extension of the area of reliance by which we enrich our subsidiary knowledge of things is an extension of the kind of knowledge we usually have of our body; it is indeed an extension of our bodily existence to include things outside it. To acquire new subsidiary knowledge is to enlarge and modify our intellectual being by assimilating the things we learn to rely on. Alternatively, we may describe the same process as an act of pouring ourselves into these things by relying on them.

Such ways of acquiring knowledge may sound strange, but then we are dealing with a kind of knowledge which, though familiar enough to us all, seems never to have been clearly identified by students of the theory of knowledge. Hitherto recognized processes for acquiring knowledge, whether by experience or deduction, apply only to the knowledge of things we are attending to and not to what we know of things by relying on our awareness of them in the process of attending to something else. I shall continue, therefore, my account of the way such knowledge is acquired and held, however curious this account may sound at first hearing.

I have said that when we rely on our awareness of some things for attending to something else, we assimilate these things to our body. In this sense, then, subsidiary knowledge is held by indwelling. We comprehend the particulars of a whole in terms of the whole by dwelling in the particulars; or, in other words, we grasp the joint meaning of the particulars by dwelling in them.

All my earlier examples of comprehension can be seen to illustrate this conclusion. To diagnose a disease is to grasp the joint meaning of its symptoms, many of which we could not specify; so we know these particulars only by relying on them as clues. Indwelling comes out more evidently when applied to the skillful testing of an object, or to any other feat of dexterous handling. Here we literally dwell in the innumerable muscular acts which contribute to our purpose, and this purpose is their joint meaning. But indwelling is perhaps most vivid in man's use of language. Human intelligence, which surpasses that of animals, comes into existence only by grasping the meaning and mastering the use of language. Little of our mind lives in our natural body; a truly human intellect dwells in us only when our lips shape words and our eyes read print. The intellectual difference between a naked pigmy of central Africa and a member of the French Academy is grounded in the cultural equipment by which Paris surpasses the African jungle. The French academician's superior mind is formed and dwells in his intelligent use of this superior equipment.

At this point we see before us a way of knowing a human being in the fullness of his dignity through recognizing in him the same powers of understanding by which we are understanding him. But let us look first at the way comprehension is achieved–comprehension as understood by my examples. More often than not we comprehend things in a flash. But it is more illuminating to think of the way we struggle from a puzzled incomprehension of a state of affairs toward its real meaning. The success of such efforts demonstrates man's capacity for knowing the presence of a hidden reality accessible to his understanding. This capacity is at work in all our knowing, from the dawn of discovery to the holding of established truth. Our active foreknowledge of an unknown reality is the right motive and guide of knowing in all our mental endeavors. Formal processes of inference cannot thrust toward the truth, for they have neither passion nor purpose. All explicit forms of reasoning, whether deductive or inductive, are impotent in themselves; they can operate only as the intellectual tools of man's tacit powers reaching toward the hidden meaning of things.

Plato argued that the task of solving a problem is logically absurd and therefore impossible. For if we already know the solution, there is no occasion to search for it; while if we do not know it, we cannot search for it either, since we do not know then what we are looking for. The task of solving a problem must indeed appear self-contradictory unless we admit that we can possess true intimations of the unknown. This is what Plato's argument proves, namely, that every advance in understanding is moved and guided by our power for seeing the presence of some hidden comprehensive entity behind yet incomprehensible clues pointing increasingly toward this yet unknown entity.

When a student is taught how to identify a disease or a biological specimen, his confidence in the hidden coherence of a puzzling state of affairs is guided by an external aid. For example, when the psychiatrist I mentioned said to his students that they would learn to recognize in practice the characteristic appearance of an epileptic seizure, he meant that they would learn to do so by accepting *his own diagnosis* of such cases and trying to understand what he based it on. All practical teaching, the teaching of comprehension in all the senses of the term, is based on authority. The student must be confident that his master understands what he is trying to teach him and that he, the student, will eventually succeed in his turn in understanding the meaning of the things which are being explained to him.

But whether our confidence in the powers of our comprehension arises spontaneously from the depths of our inquiring minds or leans on our trust in the judgment of our teachers, it is always an act of hope akin to the dynamism of all human faith. Tillich (1957) says that "that which is meant by an act of faith cannot be approached in any other way than through an act of faith" (pp. 10-11). And this holds here too. There is no other way of approaching a hidden meaning than by entrusting ourselves to our intimations of its yet unseen presence. And such intimations are the only path toward enlarging and upholding our intellectual mastery over our surroundings.

Tillich says that his dynamic conception of faith "is the result of conceptual analysis, both of the objective and subjective side of faith." This is precisely what I claim for my derivation of the dynamic conception of knowing. It is derived in the last resort from our realization of the two kinds of knowledge which combine into the understanding of a comprehensive entity. Our reliance on our awareness of the particulars is the personal; our knowledge of the entity, the objective element of knowing.

The dynamic impulse by which we acquire understanding is only reduced and never lost when we hold knowledge acquired and established by this impulse. The same impulse sustains our conviction for dwelling in this knowledge and for developing our thoughts within its framework. Live knowledge is a perpetual source of new surmises, an inexhaustible mine of still hidden implications. The death of Max von Laue should remind us that his discovery of the diffraction of X-rays by crystals was universally acclaimed as an amazing confirmation of the existing theory of crystals and X-rays. In a like manner, Dalton's atomic theory was an amazing confirmation of Boyle's speculation on the structure of crystals, which itself was a development of ideas originating with Lucretius and Epicurus. And Dalton's theory was amazingly confirmed in its turn by the experiments of J. J. Thompson 90 years later. To hold knowledge is indeed always a commitment to indeterminate implications, for human knowledge is but an intimation of reality, and we can never quite tell in what new way reality may yet manifest itself. It is external to us; it is objective; and so its future manifestations can never be completely under our intellectual control.

So all true knowledge is inherently hazardous, just as all true faith is a leap into the unknown. Knowing includes its own uncertainty as an integral part of it, just as, according to Tillich, all faith necessarily includes its own dubiety.

The traditional division between faith and reason, or faith and science (which Tillich, too, erroneously reaffirms), reflects the assumption that reason and science proceed by explicit rules of logical deduction or inductive generalization. But I have shown that these operations are impotent by themselves, and I could add that they cannot even be strictly defined by themselves. To know is to understand, and explicit logical processes are effective only as tools in search of the solution of a problem, commitment by which we expand our understanding and continue to hold the result. They have no meaning except within this informal dynamic context. Once this is recognized, the contrast between faith and reason dissolves, and the close similarity of this structure emerges in its place.

Admittedly, religious conversion commits our whole person and changes our whole being in a way that an expansion of natural knowledge does not do. But once the dynamics of knowing are recognized as the dominant principle of knowledge, the difference appears only as one of degree. For—as we have seen—all extension of comprehension involves an expansion of ourselves into a new dwelling place, of which we assimilate the framework by relying on it as we do on our own body. Indeed, the whole intellectual being of man comes into existence in this very manner by absorbing the language and the cultural heritage in which he is brought up. The amazing deployment of the infant mind is stirred on by a veritable blaze of its confidence, in surmising the hidden meanings of speech and other adult behavior, and so eventually grasping their meanings. Moreover, the child's dynamic intellectual progress has its closely similar counterpart on the highest leveis of man's creative achievement—and the structure of both these processes resembles in its turn that of the self-transformation entailed in a religious conversion.

But perhaps the deepest division between reason and faith arises from the urge toward objectivity which tends to destroy the I-Thou commitment of the religious world view and establish a panorama of I-It relations in its place. Has not the modern positivist outlook exercised its pressure even on the purely secular studies of the human mind, as well as of human affairs whether past or present, in favor of a mechanical conception of man which represents him as a bundle of appetites, or as a mechanical toy, or as a passive product of social circumstances?

But this too is the outcome of the obsessive limitation of knowledge to the results of explicit inferences. Persons can be identified as comprehensive entities only by relying on our awareness of numberless particulars, most of which we could never specify in themselves. This is the same process by which we diagnose an elusive illness or read a printed page. Just as we assimilate the symptoms of a disease by attending to their meaning, so we assimilate the workings of another man's mind by attending to his mind. In this sense we may be said to know his mind by dwelling in its manifestations. Such is the structure of empathy (which I prefer to call conviviality), which alone can establish a knowledge of other minds—and even of the simplest living beings.

Behaviorism tries to replace convivial knowledge by I-It observations of the particulars by which the mind of an individual manifests itself and tries to relate these particulars to each other by a process of explicit inference. But since most of the particulars in question cannot be observed in themselves at all and, in any case, their relation cannot be explicitly stated, the enterprise ends up by replacing its original subject matter with a grotesque simulacrum of it in which the mind itself is missing. The kind of knowledge which I am vindicating here, and which I call *personal knowledge,* casts aside these absurdities of the current scientific approach and reconciles the process of knowing with the acts of addressing another person. In doing so, it establishes a continuous ascent from our less personal knowing of inanimate matter to our convivial knowing of living beings and beyond this to the knowing of our responsible fellow men. Such, I believe, is the true transition from science to the humanities and also from our knowing the laws of nature to our knowing the person of God.

But is the person we may know in this manner not floating vaguely above his own bodily substance, outside of which he actually cannot exist at all? The answer to this question will reveal a surprising affinity between my conception of personhood and a central doctrine of Christianity.

I have said that the mind of a person is a comprehensive entity which is not specifiable in terms of its constituent particulars; but this is not to say that it can exist apart or outside of these particulars. The meaning of a printed page cannot be specified in terms of a chemical analysis of its ink and paper, but neither can its meaning be conveyed without the use of a physical medium, such as ink and paper. Though the laws of physics and chemistry apply to the particles of the body, they do not determine the manifestations of the mind; their function is to offer an opportunity–an admittedly limited and precarious opportunity–for the mind to live and manifest itself. Our sense organs, our brain, the whole infinitely complex interplay of our organism offer to the mind the instruments for exercising its intelligence and judgment, and, at the same time, they restrict the scope of this enterprise, deflecting it by delusions, obstructing it by sickness, and terminating it by death.

The knowing of comprehensive entities thus establishes a series of ascending levels of existence. The relationship I have just outlined obtains throughout between succeeding levels of this hierarchy. The existence of a higher principle is always rooted in the inferior levels governed by less comprehensive principles. Within this lower medium, and by virtue of it, the higher principle can operate widely but not unconditionally, its range being restricted and its every action tainted by the very medium on which it has to rely for exercising its powers.

We see, then, that as the rising levels of existence were created by successive stages of evolution, each new level achieved higher powers entramelled by new possibilities of corruption. The primeval matrix of life was inanimate and deathless–subject to neither failure nor suffering. From it have emerged levels of biotic existence liable to malformation and disease and, at higher stages, prone also to illusion, to error, to neurotic affliction–finally producing in man, in addition to all these liabilities, an ingrained propensity to do evil. Such is the necessary condition of a morally responsible being, grafted on a bestiality through which alone it can exercise its own powers.

Such is the inescapable predicament of man which theology has called his fallen nature. Our vision of redemption is the converse of this predicament. It is the vision of a man set free from this bondage. Such a man would be God incarnate; he would suffer and die as a man, and yet by this very suffering and death he would prove himself divinely free from evil. This is the event, whether historical or mythical, which shattered the framework of Greek rationalism and has set for all time the hopes and obligations of man far beyond the horizons of Greek philosophy.

I have mentioned divinity and the possibility of knowing God. These subjects lie outside my argument. But my conception of knowing opens the way to them. Knowing, as a dynamic force of comprehension, uncovers at each step a new hidden meaning. It reveals a universe of comprehensive entities which represent the meaning of their largely unspecifiable particulars. A universe constructed as an ascending hierarchy of meaning and excellence is very different from the picture of a chance collocation of atoms to which the examination of the universe by explicit modes of inference leads us. The vision of such a hierarchy inevitably sweeps on to envisage the meaning of the universe as a whole. Thus natural knowing expands continuously into knowledge of the supernatural.

The very act of scientific discovery offers a paradigm of this transition. I have described it as a passionate pursuit of a hidden meaning, guided by intensely personal intimations of this yet unexposed reality. The intrinsic hazards of such efforts are of its essence; discovery is defined as an advancement of knowledge that cannot be achieved by any, however diligent, application of explicit modes of inference. Yet the discoverer must labor night and day. For though no labor can make a discovery, no discovery can be made without intense, absorbing, devoted labor. Here we have a paradigm of the Pauline scheme of faith, works, and grace. The discoverer works in the belief that his labors will prepare his mind for receiving a truth from sources over which he has no control. I regard the Pauline scheme therefore as the only adequate conception of scientific discovery.

Such is, in bold outline, my program for reconsidering the conception of knowledge and restoring thereby the harmony between faith and reason. Few of the clues which are guiding me today were available to the Scholastics. The modes of reasoning which they relied on were inadequate; their knowledge of nature was tenuous and often spurious. Moreover, the faith they wanted to prove rational was cast into excessively rigid and detailed formulae, presenting intractable and sometimes even absurd problems to the reasoning mind.

Even so, though their enterprise collapsed, it left great monuments behind it, and I believe that we are today in an infinitely better position to renew their basic endeavor. The present need for it could not be more pressing.

9

ON THE MODERN MIND

A few years ago I gave a talk on the modern mind to the Medical Section of the British Psychological Society, by which my audience seemed to be disappointed, and my chairman said so. They had hoped for something more substantial, he said, "something to get one's teeth into." From his point of view he was right. For I spoke of the modern mind as a body of ideas having their origin in thought, while in his profession he was used to regarding ideas as the rationalization of drives, of guilt, or anxiety, or aggression, or insecurity. Such a view is widespread. Here too it may be felt that I am not dealing with the tangible forces determining the mind. When I go on ignoring infantile traumas, broken homes, industrialization, many may feel lost in a world of shadows.

In a way I should welcome such opposition, as it would help to establish my first point, which is that the modern mind distrusts intangible things and looks behind them for tangible matters, on which it relies for understanding the world. We are a tough-minded generation.

My second point makes a curious pair with the first. For it is that in spite of our tough theories, our society is more humane than any that had existed before. And if our terrible wars and revolutions are cited against this, I would reply in the words of Paul Tillich: "If ever in history there was a time when human objectives supported by an infinite amount of good will heaped disaster upon disaster on mankind, it is the twentieth century." I would

First published in *Encounter*, 15(May): 12-20, 1965.

say that ideals, the genuineness of which our skepticism has taught us to question, have in fact swayed our time and by their power have almost shattered our civilization.

I would go further and add that, if our skepticism itself goes to extremes, it does so in the pursuit of a moral purpose, namely of a relentless intellectual honesty. The two conflicting ideas of our age–its skepticism and its moral passions–are indeed locked in a curious struggle in which they may combine and reinforce each other. This is a strange story.

The beginnings of modern skepticism go back to ancient Greece, but its present overpowering strength is the sequel of the Copernican revolution. Copernicus ousted man from his central position in the universe and destroyed the theological cosmos. The heavenly sphere of divine perfection toward which, from his fallen sublunar existence, man was bound to strive, was dissolved in a space without limits, without shape or center.

And monotony in space was extended into monotony in depth by the atomic theory of matter. Galileo's mechanics, amplified by Newton, gave new life to the theory that all things are ultimately composed of masses in motion. Atomic particles alone were real and all phenomena were merely appearances of this ultimate reality. Man himself was but a chance collocation of atoms, without purpose or meaning.

Yet the new fellow feeling, the other master idea of modern man, standing opposite to skepticism, was born indirectly from skepticism. For it was the attack of skepticism on the Christian churches that released the moral ideals of Christianity from a striving for individual salvation and directed our moral conscience instead to the betterment of human society. The imagination of the new rationalism was soon to be aflame with aspirations for a higher condition of man and society.

Throughout all previous ages men had accepted existing custom and law as the foundations of society. There had been changes and some great reforms, but never before had the deliberate contriving of unlimited social improvement been elevated to a dominant principle. The first government to adopt this principle was that established by the French Revolution; and so the turn of the eighteenth century became the dividing line between the immemorial expanse of virtually static societies and the following brief period in which a passionate hope for a better future became a dominant force in public life.

Mechanical Philosophy

Scientific skepticism smoothly cooperated at first with the new passion for social betterment. Battling for freedom of thought against established authority, skepticism cleared the way for political freedom and humanitarian reforms. Scientific rationalism brought social and moral progress that has improved almost every human relationship in Western civilization. The new rationalism has been, up to our own days, the chief guide toward intellectual, moral, and social advances.

But troubles developed and became serious in our own century. The demand that all things must be explained by the laws of physics and chemistry became more insistent and more disturbing. A sharpening of skepticism to the point of questioning the very existence of intangible things led to absurd conclusions.

I shall try to restore here our acceptance of higher forms of being and to show how we *can* know and *do* know these less tangible levels of existence. I will then try to bring into view the second master idea of our age and show how this idea of unlimited progress, intensified to perfectionism, has combined with our sharpened skepticism to produce the perilous state of the modern mind. I shall speak of the disasters of our age and finally tell of signs pointing toward a recovery of our basic ideals.

The mechanical philosophy of Galileo was more fully stated by Laplace when he defined a Universal Knowledge of the World. He pointed out that from today's topography of the ultimate particles of the world (which would include their velocities and the forces acting between them) we could calculate any future topography of the same particles, and he claimed that this would give us a knowledge of all things to come, to the very end of time. It has been objected that such predictions contradict the exercise of free will, but this had only the effect of calling into question our possession of free will. Indeed, to bring up this particular difficulty of free will is to overlook the more massive fact that a Laplacean atomic topography would tell us virtually nothing that is of interest to us. It would give us the total energy of any particular region in the universe, but we could not even make out whether things in that region had any definite temperature, and if so what that temperature was.

To fathom the depth of such ignorance parading as Universal Knowledge, imagine yourself deprived of all your previous experience and presented in its place with a Laplacean topography of the universe. Though you were endowed with an unlimited ability for mechanical computations, you would search in vain to calculate something worth knowing. For what you would want to know are things seen and felt, things heard and smelt, and the laws of mechanics cannot derive such knowledge from a topography of atomic particles. Only the action of our sentient self, responding to the atoms impinging upon our senses, can supply such information.

But even granting, for the sake of argument, our powers of sentience, and forgetting also that an atomic topography cannot define temperature, we could still get no further than to derive the laws of physics and chemistry, and this would not enable us to recognize living and sentient beings. In saying this, I contradict the claims of biologists who affirm that they are explaining life in terms of physics and chemistry. But the fact is that they do nothing of the kind. The purpose which biology actually pursues, and by which it achieves its triumphs, consists in explaining the functions of living beings in terms of a mechanism *founded* on the laws of physics and chemistry, yet *not explicable by* these laws.

We can make this clear by showing that no mechanism, not even the simplest machine, can be explained in terms of physics and chemistry. Let me choose as an example of a machine the watch I wear on my wrist. My watch tells the time. It is kept going by the mainspring, uncoiling under the control of the hairspring and balance wheel, and thus it turns the hands which tell the time. Such are the operational principles of a watch, the principles which define its construction and working. It is these principles that cannot be defined by the laws of inanimate nature. For no part of a watch is formed by the natural equilibration of matter. Each is artificially shaped and connected to perform its function.

Physics and chemistry cannot reveal the practical principles embodied in a machine, any more than the physical chemical testing of a printed page can tell the content of its text.

But how can we embody any structural or operational principle in a piece of inanimate matter without interfering with the laws of inanimate matter? To answer this question we must realize that no inanimate object is ever fully determined by the laws of physics and chemistry. Laplace himself had to assume for his speculations on future atomic topographics an initial atomic topography which was not derived from atomic mechanics. *The laws of physics and chemistry can likewise be applied only to a given set of initial conditions.*

This is true for any general principle that applies to experience. It must leave indeterminate a certain range of circumstances in which it can apply, and any particular application of such a principle requires that these circumstances be fixed by some agency not under the control of that principle. This is well known for the laws of physics. The conditions which have to be fixed by some external agency are called here the *boundary conditions* of the system to which these laws are applied. It is on these boundary conditions that the shaping of a piece of metal into a machine takes effect. *Machines are systems, in which the boundary conditions left open by physics and chemistry are controlled by certain structural and operational principles;* and hence machines cannot be described in terms of physics and chemistry.

And what is true of machines is, of course, equally true of the machinelike functions of living beings. Such functions are determined by structural and operational principles which control the boundary conditions left open by physics and chemistry. Living conditions cannot, therefore, be described in terms of physics and chemistry.

Thus the material of a machine is under the control of these two independent principles. The role of these two is very different. If the laws of physics and chemistry were suspended for a moment, all machines would stop working; their operational principles rely for their performance on these laws. This of course does not hold in reverse. Pulverize a machine and its fragments will continue to obey the laws of physics and chemistry. The wrecking of the operational structure does not affect these laws, for they apply to the material of the machine in itself even when split into isolated bits of matter.

A machine or a machinelike functioning living being can be said, therefore, to comprise two levels. There is an upper, comprehensive level embodying the operational principles of the system and a lower, more primitive level, controlled by the laws of physics and chemistry. The lower level is formed by the unorganized mass, the higher level by the principle that controls its organization. In other words, we have a lower level of isolated parts and a higher level of the functional whole formed by the parts. This higher level represents then the joint "meaning" of the parts.

We see here the beginnings of a hierarchy in which the distinction between things essentially higher and essentially lower is restored.[1]

We can generalize the two-level structure of living beings and machines to the playing of a game of chess. The conduct of such a game is an entity controlled by a stratagem and the stratagem relies on the observance of the rules of chess. This relation does not hold in reverse, for the rules of chess leave open an infinite range of stratagems. Chess moves are therefore meaningless by themselves; their meaning lies in serving jointly the performance of a stratagem.

Sequence of Levels

All these relations become clearer in the case of a skill which comprises a number of levels in the form of a hierarchy. The production of a literary composition, for example a speech, includes five levels. The first level, lowest of all, is the production of a voice; the second, the utterance of words; the third, the joining of words to make sentences; the fourth, the working of sentences into a style; the fifth, and highest, the composition of the text.

The principles of each level operate under the control of the next higher level. The voice you produce is shaped into words by a vocabulary; a given vocabulary is shaped into sentences in accordance with a grammar; and the sentences are fitted into a style,

[1]The argument of this section was first developed in this form in Polanyi (1962).

which in its turn is made to convey the ideas of the composition. Thus each level is subject to dual control; first, by the laws that apply to its elements in themselves, and second, by the laws that control the comprehensive entity formed by them.

Such multiple control is made possible again by the fact that the principles governing the isolated particulars of a lower level leave indeterminate their boundary conditions, to be controlled by a higher principle. Voice production leaves largely open the combination of sounds into words, which is controlled by a vocabulary. Next, a vocabulary leaves largely open the combination of words to form sentences, which is controlled by grammar; and so the sequence goes on. Consequently, the operations of a higher level cannot be accounted for by the laws governing its particulars forming the next lower level. You cannot derive a vocabulary from phonetics; you cannot derive grammar from a vocabulary; a correct use of grammar does not account for good style; and a good style does not supply the content of a piece of prose.

A glance at the functions of living beings assures us that they consist in a whole sequence of levels forming such a hierarchy. The lowest level is controlled by the laws of inanimate nature and the higher levels control throughout the boundary conditions left open by the laws of the inanimate. The lowest functions of life are those called vegetative; these vegetative functions, sustaining life at its lowest level, leave open—in both plants and animals—the higher functions of growth, and leave open in animals also the operations of muscular action; next in turn, the principles governing muscular action in animals leave open the integration of such action to innate patterns of behavior; and again such patterns are open in their turn to be shaped by intelligence; while the working of intelligence itself can be made to serve in man the still higher principles of a responsible choice.

We have thus a sequence of rising levels, each higher one controlling the boundaries of the one below it and embodying thereby the joint meaning of the particulars situated on the lower level. The meaning of each successive rising level thus becomes richer at each stage and reaches the fullest measure of meaning at the top. We can see then why the Universal Knowledge of Laplace, or a physicochemical topography of the world, is virtually meaningless. *All meaning lies in higher levels of reality that are not reducible to the laws by which the ultimate particulars of the universe are controlled.*

The world view of Galileo, accepted since the Copernican revolution, proves to be fundamentally misleading. *What is most tangible has the least meaning and it is perverse then to identify the tangible with the real.* For to regard a meaningless substratum as the ultimate reality of all things must lead to the conclusion that all things are meaningless. And we can avoid this conclusion only if we acknowledge instead that deepest reality is possessed by higher things that are least tangible.

Positivism

This might seem to take us back to the medieval conception, which interpreted the particulars of the world in the light of its major comprehensive meaning which understood the parts as manifestations of the whole and not the other way round. But this would be going too fast. It is not enough to show that there is room for living functions and other higher principles in the boundaries left open by the laws governing inanimate nature. For we cannot claim the existence of essentially higher levels so long as our very identification of them is called into question by a positivistic empiricism. To this objection I shall now turn.

There is no precise theory of positivistic empiricism, but its present practice is clear in some important cases. It denies that we can know more than tangible, external facts. Take the consciousness of a human being. Consciousness, we are told, is not a tangible fact and we must therefore hesitate to attribute consciousness to any living being, animal or man.

It would seem impossible that neurophysiologists, let alone psychologists, should deny the existence of consciousness, which is a major part of their subject matter. Can one study perception without referring to what people see? Or the localization of emotional centers in the brain without referring to what the subjects feel? Yet a distinguished neurophysiologist like D. O. Hebb (1954) has urged scientists to assume that consciousness does not exist, even though such a hypothesis might eventually prove false. Nor is this an isolated instance. The psychiatrist L. S. Kubie (1954), speaking on the same scientific occasion, declared that a "working concept" of consciousness was indispensable to psychology, and went on to say: "sometimes we are explicit and frank about this. Sometimes we fool ourselves about it. Many workers have attempted to avoid using the word because of its traditional connotations, which have had a somewhat mystical, imponderable, non-scientific, philosophic and/or theological flavour" (pp. 445-446). Kubie's words show what is happening here. Scientists who urge us to assume that consciousness does not exist do not believe this themselves. It would be absurd to suppose that Hebb wants neurophysiologists to assume that all their subjects are unconscious. He merely wants them to describe their findings as if consciousness did not exist.

This is the program of behaviorism. It sets out, for example, to eliminate all references to the human mind, by substituting for the mind the sound of human speech when telling about a state of mind. Such an inquiry refuses to observe that a man is in pain and it can acknowledge only that he complains of pain. The fact that this view wipes out the purpose of medicine–as the alleviator of human suffering–is disregarded. Behaviorism could describe medicine only as a process for eliminating complaints of pain, even though complaints can be more effectively silenced without medicine. The very conception of compassion is denied and torture is theoretically given free rein.

None of this is intended, or even remotely approved, by behaviorists who call into question the existence of consciousness. It is clear, therefore, that they do not mean what they say when they urge us to doubt or disregard, or at least avoid mentioning, the existence of consciousness. They seem to take pride, as scientists, in professing something that laymen would find absurd. They feel themselves then as successors to the Copernicans who forced laymen to see our earth, the very ground of fixity, hurtling around an immobile sun.

Such fooling of ourselves is widely admitted in biology. Everyone knows that you cannot inquire into the functions of living organisms without referring to the purpose served by these functions, and by the organs and processes which perform these functions. Yet we must pretend that all such teleological explanations are merely provisional. The story goes round among biologists that teleology is a woman of easy virtue, whom the biologist disowns in public, but lives with in private.

The practice of science can be sound even when it is conducted in the name of false principles. For biologists to deny their use of teleological reasoning is quite harmless. It is even possible that some valuable research *must* be based on absurd assumptions. Think of the recent exploration of various parts of the brain by electrodes of microscopic size, which showed the nervous system operating as a machine. This splendid inquiry would be hampered by keeping in mind the fact that the assumption of the whole nervous system operating as an insentient automaton is nonsensical. Neurologists may be right, therefore, in ignoring the absurdity of the idea underlying their work.

The situation reminds one of the theological map-makers of the Middle Ages. In the sixth century a great traveller and merchant, called Cosmas, turned monk and then launched an attack against Greco-Roman geography on the grounds that it contradicted the text of the Bible. He produced in its stead an image of the world in the shape of the tabernacle of Moses. It looked like an old-fashioned trunk, with its lid as the heavenly firmament. Other absurd theological maps were current in the Middle Ages until the fifteenth century, even while sailors' maps of remarkable precision were used to travel the seas of Europe. Cosmas himself would not have relied for his travels to India on his tabernacle as the map of the world. But having turned monk, he found this image professionally illuminating.[2]

The official theories of psychology and biology also give professional satisfaction, even though nobody can believe in them. Yet I think it would be better to stick to the obvious truth, if this can be done with a good philosophic conscience, as I think it can.

We are told that the consciousness of another person is not directly observed but merely inferred from external facts, and that a strict empiricism prefers to acknowledge only facts that are directly observed. But nothing is ever observed except by the aid of intelligent transactions which integrate a great number of impacts made on our several senses, along with the internal responses

[2]See Brown (1949): "After tearing down the writings of Greek infidels such as Plato, Aristotle, Eudoxus, and Ptolemy, he [Cosmas] proceeds to construct his own cosmography based on the Scriptures and the writings of the Holy Father." The *portolano* or harbor-finding charts accurately described the coastline of the Mediterranean.

evoked by these impacts within our own body. What we see and hear depends in a thousand ways on the preparedness of our own mind and on our intelligent participation in making out what it is that we see and hear.

Suppose I look at my right hand. I recognize its area by its closed contours. But if that were all, my hand, when moved about, would keep changing its color, its shape, and its size. The experience of my hand as a solid object, having definite properties, would never arise. I see it as such by integrating a host of rapidly changing clues both in the field of vision and inside my eyes and some still deeper in my body. By my powers of integration I see thousands of changing clues jointly, as one single unchanging object moving about at different distances, viewed from different angles, under variable illuminations. And this is exactly what happens when I observe a face full of anger and menace; I see it by exactly the same kind of integration. We cannot reasonably reject our observation of anger and menace on the ground that it requires an act of integration, unless we refuse to observe anything at all.

We can deepen this result by exploring the process of integration a little further. Suppose I look at an object, for example, my own finger, through a pinhole in a sheet of paper, or better still, through a blackened tube; if I do this and then move my finger back and forth I see it swelling as it approaches my eye. The moving object has lost some of its solidity, for it now lacks confirmation by the clues that normally contribute to its image from the periphery of the visual field.

Notice here how many of the clues we integrate so successfully to the sight of an object are not known to us in themselves. Many of them cannot be sensed at all; the contractions of our eye muscles, for example, can never be experienced as such and we are aware of them only in the way they make us see the object that we are looking at. Other clues, like those we cut out by a pinhole, we do sense, but only from the corner of our eye. We do not attend to these either, *but rely on our awareness of them for attending to the coherent entity to which they contribute.*

Modern philosophers have argued that we can have no evidence for inferring the existence of other minds, and this would be true if we had to rely on an explicit process of inference. But that is not the case. We integrate the particulars of a physiognomy in the same way as we integrate the clues, or parts, of any other perception, namely by fusing the clues or parts as presented to our senses into a meaningful way of perceiving them. We may call this a *tacit process* of inference by contrast to an explicit process of inference as defined by logic today.

The new element I have introduced here into the conception of knowing is the way we know clues by relying on our awareness of them for attending to that to which they point or, more generally, the way we know things by relying on our awareness of them for attending to something else, which is the coherent entity to which they contribute. We may link this now with a certain experience we all have of things we know almost exclusively by relying on them for attending to something else. *Our own body is an assembly of such things.* For we hardly ever attend to our body as we attend to an external object, while we continually rely on it as a means for observing objects outside and for manipulating these objects for our own purposes. *We may identify, therefore, our knowing of something by attending to something else with the kind of knowledge we have of our own body by dwelling in it.* In other words, we may say that when we rely on our awareness of some things for attending to other things, we have assimilated these things to our body. We may say, for example, that we know the clues of perception by dwelling in them, when we attend to that which they jointly indicate; and that we see the parts of a whole forming the whole by dwelling in the parts. We thus arrive at the conception of *knowing by indwelling.*

Indwelling operates on all levels of reality. But when we know living things, our indwelling enters into an especially intimate relation to that which it knows. A lion pouncing on the back of an antelope coordinates its own observations and actions in a highly complex and accurate way. The naturalist watching the lion mentally integrates these coordinated elements into the conception of the lion hunting its prey. Other vital coordinations, like embryonic development, are much slower than this, but no less rich in coordinated details; the study of physiological functions fills many volumes and the coordinations performed by human intelligence are unlimited. But the perception of living beings consists throughout in mentally duplicating the active coordinations performed by their living functions.

We can see now how we know another man's mind and share his mental life. We understand, for example, a man's skillful performance by mentally combining its several movements into their joint pattern. Chess players enter a master's mind by rehearsing the games he has played. Knowing a man's mind is then to experience the joint meaning of his actions by dwelling in them from outside. This is how we get to feel another man's consciousness, to share his pain and pity him. Knowing life is always a sharing of life, but to know another person is to share his life as an equal partner. When we study inanimate matter or the lower organisms, we stand to these in an I-It relation, but as we gradually rise to the study of man we arrive at an I-Thou relation to him. We enter into mutual understanding with him.

Here, then, is a theory of knowledge which tells us how we can both know and experience the higher intangible levels of existence, which a positivistic empiricism refuses to recognize.

I shall now pass on to some large questions of our culture, by facing the challenge that a positivistic empiricism presents to the existence of moral principles. A textbook of sociology (Johnson, 1960) opens with a formal statement of its principles in four points. The fourth of these principles declares that sociology is "*unethical;* that is, sociologists do not ask whether particular social actions are good or bad; they seek merely to explain them." Some sociologists would seek to qualify this principle, but very few effectively do so. It is predominantly accepted and cherished as securing the scientific character of sociology.

Let us face what is implied in this principle. To assume that you can explain an action without regard to whether it is good or bad is to assume that moral motives play no part in it. To extend this assumption to all social action is to deny the very existence of genuine moral motives in men. When I protest against such doctrines, I am assured that the sociologists who teach this moral nihilism are themselves men of high moral principles, supporting noble causes in public life. This is thought to put the matter right. It is considered quite in order that we should teach absurd views that we do not believe because we think that they are scientific.[3]

I admit that most students will uphold their moral convictions regardless of being taught that these are without foundation. They may even respond to the social perfectionism of our age, and make high moral demands on society. Some may never feel this internal contradiction; in others it may cause confusion, a reduction of respect for their own lives. We can assess the possible consequences arising from this situation by turning to the ideas of writers who have worked out these contradictions of the modern mind in literature and political thought. Literature and politics are the mythology of our age and the school of our imagination.

The tension between a positivist skepticism and a modern moral perfectionism has indeed erupted with vast consequences in our days. It erupted in two directions, toward art and philosophy, and toward politics. The first was a move toward extreme individualism, the second, on the contrary, toward modern totalitarianism. These two movements may appear diametrically opposed, yet they are but two alternative solutions of the same

[3]In the Foreword to Johnson (1960), Robert K. Merton writes: "With this book Mr. Johnson joins the small circle of... masters of sociological writing."

Freedom from value judgment is maintained throughout the book. Cruelty to Negroes in the latter half of the last century is explained by the use of the Negro as a scapegoat. Victimizing the Negro deflects feelings of frustration from causing social disruption. "Thus the Negroes were victims of a heightened need of national unity in the face of external problems.... National unity and sectional unity were achieved partly at the expense of the Negro" (p. 602).

The social functions of scapegoating have been repeatedly analyzed during the past decades, for example by Kluckhohn and Leighton (1946, pp. 176-177): "... Navahos 'take out' on witches by word and by deed the hostility which they feel against their relatives, against whites, against the hazards of life itself.... The killing of witches is characteristically brutal.... Witches in other words are scapegoats." "... there is no doubt that witchcraft is Navaho culture's principal answer to the problem that every society faces, how to satisfy hate and yet keep the core of society solid.... The people blame their troubles upon 'witches' instead of upon 'Jews' or 'niggers'."

This theory leads Johnson to comment on the further development of the Negro's position in the United States as follows: "Technically, perhaps, the suffering of the Negro is no more dysfunctional than the loss of men in a victorious battle. Everything, including the integration of social systems, is achieved at a cost." Nevertheless, we are told, the necessity of using Negro talent and of placating African states has led to concessions to the Negro.

Would it not appear, then, that if Lincoln had but known of the social functions of "scapegoating," he might have introduced Negro-baiting in the North and thus united the nation at a much lesser cost than by a civil war?

equation which required *the joint satisfaction of a belief in moral perfection with a complete denial of moral motives.*

I shall start with the individualist solution of this equation. A man looking at the world with complete skepticism can see no grounds for moral authority or transcendent moral obligations; there may seem to be no scope then for his moral perfectionism. Yet he can satisfy it by turning his skepticism against existing society, denouncing its morality as shoddy, artificial, hypocritical, and a mere mask for lust and exploitation. Though such a combination of his moral skepticism with his moral indignation is inconsistent, the two are in fact fused together by their joint attack on the same target. The result is a moral hatred of existing society and the alienation of the modern intellectual.

The effect on his inner life goes deep. His skepticism-with-perfectionism scorns any expression of his own traditional morality; it despises it as banal, second-hand, hypocritical. Divided against himself, he seeks an identity safe against self-doubt. Having condemned the distinction between good and evil as dishonest, he can find pride in the honesty of his condemnation. Since ordinary decent behavior can never be safe against the suspicion of sheer conformity or downright hypocrisy, only an absolutely amoral, meaningless act can assure man of complete authenticity. All the moral fervor which scientific skepticism has released from religious control and then rendered homeless by discrediting its ideals returns then to imbue an amoral authenticity with intense moral approval. This is how absolute self-assertion, fantasies of gratuitous crime and perversity, self-hatred, and despair are aroused as defenses against a nagging suspicion of one's own honesty.

This theme has prevailed in Continental thought since Dostoevsky, a century ago, first described murder as an experiment in moral skepticism and, soon after, Nietzsche repudiated all traditional conceptions of good and evil as hypocritical. About the same time Rimbaud launched a great poet's imagination into a world of disordered sensualism, and he was followed in the next generation by Gide, who showed that perversion and gratuitous crime could be marks of moral authenticity. Today we have a whole literature, much of it of high quality, in which absurdity and a somber, fantastic obscenity are lived as tokens of unflinching honesty.

These are individualistic solutions of the conflict between skepticism and perfectionism. They unite the two opposites in a moral nihilism charged with moral fury. This paradoxical combination is new in history and deserves a new name: I have called it a *moral inversion.*

In public life moral inversion leads to totalitarianism. Marxism-Leninism is the most important movement of this kind. The Marxist revolutionary scorns any appeal to generous sentiments and scorns any appeal to the utopian image of an ideal society. His skepticism forbids him to acclaim such motives. But, though he cannot declare these high motives, they are his driving force and must be satisfied. Marxism resolves this contradiction by inventing a machine–the Marxist machine of history–which, working inside society, will bring about the destruction of capitalism and its replacement by socialism. The machine will achieve this without the aid of noble sentiments or images of social perfection. And such a mechanism, claiming to control all mental processes in society, is bound to appeal to a scientific outlook. When this mechanism also offers a safe disguise and embodiment for the utopianism which motivates its makers, its appeal becomes irresistible.

The two contradictory elements of Marxism effectively protect its teachings against criticism by alternately taking over its defense. Its moral fervor denies a hearing to any intellectual objections, while any moral scruples are contemptuously rejected as unscientific.

This combination of conflicting principles explains how Marxists can accept historical inevitability as an incentive to work and fight for bringing about the events declared to be inevitable. For since the Marxist theory is merely a disguise for utopian ideals, it can tacitly enjoin us to fight for the fulfillment of its theoretical predictions.

The harshness of the political parties charged with this task is often criticized, or else excused as the use of evil means in the service of a noble cause; but such reproach or excuse is misplaced. Marxism-Leninism denies being guided either by moral motives or utopian visions, and declares that it follows only the directives of science. The question of weighing means against ends cannot then arise. If you claim to embody a mechanism, you must behave like a machine: your unscrupulousness will be sanctioned by the morality inside the machine. A morality embodied in a machine is necessarily blind to its own handiwork and deaf to the voices of reason: it has turned fanatical.

Truth itself then becomes embodied in the machine. Whatever makes the machine run faster is said to be true. A universe of public fantasies is erected in which even its authors have lost their bearings. The very victims of faked trials are persuaded that in some sense the fantastic accusations against them are true. To think otherwise would be to forsake the revolution, which is unthinkable.

Rescue from Disaster

I have described the modern mind by the content of its ideas and have explained the emergence of the modern mind as the outcome of a process of thought which originated in the Copernican discovery and in the interplay of the ensuing intellectual revolution with the moral ideas of Christianity. This was how I explained the moral ideas of modern literature, as well as the political creeds and disasters of our age.

These great events were not due to the effects of economic circumstances nor to the early training of infants. The ideas of the Russian Revolution have spread to regions of the most varied economic structures and of equally varied customs of swaddling babies or of early toilet training. Any theory that would account for these revolutions of thought by economic or infantile traumas expresses the same errors concerning the nature of man and history which caused these disastrous revolutions. It perpetuates these errors.

A true diagnosis of our disorders should help to overcome them. My own interpretation of the modern world would do this by recognizing thought as an independent, self-governing force.

I feel supported in this by the great movements recoiling from modern totalitarian ideologies. Stalinism is passing away and we look back on its rule with growing amazement. Russians are asking insistently how those terrible things could have happened. Concluding his memoirs in 1962, Ilya Ehrenburg speaks of "all the things that lie like a stone on the hearts of people of my generation." The whole world is involved in this: we cannot trust ourselves again unless we can understand how people, so steeped in our own modern scientific outlook, could produce such an insane tyranny and support it fanatically for years on end.

The answer to this question is coming out by stages, darkly. At the 20th Congress of the Russian Communist Party, held in February 1956, Khrushchev first denounced Stalin's misdeeds in a secret speech. A few months later Polish and Hungarian writers were openly demanding freedom of thought. These men were leading Communist intellectuals who were recoiling from the theory that morality, justice and art, and truth itself, were to be identified with the interest of the party. Hungarian Communist writers solemnly repudiated the teaching that political expediency could be a criterion of the truth and "after bitter mental struggles" they vowed "that in no circumstances would they ever write lies." A few weeks later, the Hungarian people, led by these intellectuals, overthrew the Stalinist regime established by Rakosi.

This revolution was fought to gain recognition for the reality of intangible things: of truth, of justice, of moral and artistic integrity. The Bolshevik attempt, undertaken for high purposes and in the light of a sophisticated theory, to establish an empire that denied this reality, had failed. It had proved to be unbearable. I believe that this passionate recognition of a metaphysical reality, irreducible to material elements, marks a turning point: it will serve as an axiom for any future political thought.

Writers in Poland and Hungary are now trying to find a place for the morally responsible individual within the Marxian conception of history. Early manuscripts of Marx, until recently unpublished, offer some substance for this. But the reviving of some Hegelian ideas in the thought of the young Marx will not take us far.

We need a theory of knowledge which shows up the fallacy of a positivist skepticism and authorizes our knowledge of entities governed by higher principles. Any higher principle can be known only by dwelling in the particulars governed by it. Any attempt to observe a higher level of existence by a scrutiny of its several particulars must fail. We shall remain blind in theory to all that truly matters in the world so long as we do not accept indwelling as a legitimate form of knowledge.

Indwelling involves a tacit reliance on our awareness of particulars not under observation, many of them unspecifiable. We have to interiorize these and, in doing so, must change our mental existence. There is nothing definite to which we can hold fast in such an act. It is a free commitment.

But there is something imponderable for us to rely on. We have around us great truths embodied in works born of the very freedom which we are hesitating to enter. And recent history has taught that we can breathe only in the ambience of these truths and of this creative freedom. I, for one, am prepared to rely on this assurance for acquiring and upholding knowledge by embracing the world and dwelling in it.

REFERENCES

Ashby, E. (1950), The Pattern of Soviet Science. *Listener,* 42 (Mar. 30):549-550.

Bartlett, F. C. (1932), *Remembering.* Cambridge: Cambridge University Press.

Berlin, I. (1939), *Karl Marx.* London: Oxford University Press.

Blackett, P. M. S. (1947), The Magnetic Field of Massive Rotating Bodies. *Nature,* 159:658-666.

Brown, L. A. (1949), *The Story of Maps.* Boston: Little, Brown.

Burtt, E. A. (1924), *The Metaphysical Foundations of Modern Physical Science.* Humanities Press. Rev. ed., Doubleday Anchor Books, 1954.

Childe, G. (1942), *What Happened in History.* Baltimore, Md.: Pelican, 1954.

Ehrenwald, J. (1948), *Telepathy and Medical Psychology.* New York: Norton.

Einstein, A. (1905), Zur Elektrodynamik bewegter Körper. *Ann. Physik,* 17:891-921.

______ (1949), *Albert Einstein: Philosopher-Scientist,* ed. P. A. Schilpp. Evanston, Ill.: Library of Living Philosophers.

Evans-Pritchard, E. E. (1937), *Witchcraft, Oracles, and Magic among the Azande.* London: Oxford University Press.

Gamow, G. (1966), *Thirty Years That Shook Physics.* New York: Doubleday.

Haldane, J. B. S. (1949), In Defence of Genetics. *Mod. Quart.,* 4(3):194.

Hare, R. M. (1952), *The Language of Morals.* London: Oxford University Press.

Hebb, D. O. (1954), The Problem of Consciousness and Introspection. In *Brain Mechanisms and Consciousness,* ed. J. F. Delafresnaye. Oxford: Blackwell, pp. 402-421.

Held, R. (1965), Object and Effigy. In *Vision and Value, Structure in Art and Science,* Vol. 2, ed. G. Kepes. New York: Braziller.

Huesmann, L. R., & Cheng, C.-M. (1973), A Theory for the Induction of Mathematical Function. *Psychol. Rev.,* 80:126-138.

Hull, C. L. (1943), *Principles of Behavior.* New York: Appleton-Century.

Huxley, J. (1949a), *Heredity East and West.* New York: Schuman.

______ (1949b), Soviet Genetics: The Real Issue. *Nature,* 163:935-942.

Johnson, H. M. (1960), *Sociology: A Systematic Introduction.* New York: Harcourt, Brace, and World.

Kenneth, J. H. (1940), Average Gestation Period and $n\pi$. *Nature,* 146:620.

Kluckhohn, C., & Leighton, D. (1946), *The Navaho.* Cambridge: Harvard University Press.

Kubie, L. S. (1954), Psychiatric and Psychoanalytic Considerations of the Problem of Consciousness. In *Brain Mechanisms and Consciousness,* ed. J. F. Delafresnaye. Oxford: Blackwell, pp. 444-469.

Locke, J. (1692), A Third Letter for Toleration. *Works,* Vol. 6, 10th ed. London, 1801.

Mach, E. (1883), *Die Mechanik in ihrer Entwickelung.* Leipzig: Brockhaus.

Neisser, U. (1967), *Cognitive Psychology.* New York: Appleton-Century-Crofts.

Pieper, J. (1960), *Scholasticism: Personalities and Problems of Medieval Philosophy.* New York: Pantheon.

Polanyi, M. (1946), *Faith and Society.* London: Oxford University Press.

______ (1962), Tacit Knowing: Its Bearing on Some Problems of Philosophy. *Rev. Mod. Physics,* 34:601-615.

Pribram, K. H. (1971), *Languages of the Brain.* Englewood Cliffs, N. J.: Prentice-Hall.

Rapaport, D., ed. (1951), *Organization and Pathology of Thought.* New York: Columbia University Press.

Rayleigh, Baron (1947), The Surprising Amount of Energy Which Can Be Collected from Gases after the Electric Discharge Has Passed. *Proc. Royal Soc.,* 189A:296-299.

Reichenbach, H. (1953), The Philosophical Significance of the Theory of Relativity. In *Readings in the Philosophy of Science,* ed. H. Feigl & M. Brodbeck. New York: Appleton-Century-Crofts, pp. 195-211.

Richards, I. A. (1924), *Principles of Literary Criticism,* 4th ed. New York: Harcourt, Brace, 1930.

Russell, H. N., with Dugan, R. S., & Stewart, J. Q. (1945), *Astronomy: A Revision of Young's Manual of Astronomy,* rev. ed. Boston: Ginn.

Ryle, G. (1962), *Dilemmas.* London: Cambridge University Press.

Tillich, P. (1957), *Dynamics of Faith.* New York: Harper.

Urmson, J. O. (1950), On Grading. *Mind,* 59:145-169.

Watson, J. D. (1968), *The Double Helix.* New York: Atheneum.

Wertheimer, M. (1959), *Productive Thinking.* New York: Harper.

Weyl, H. (1949), *Philosophy of Mathematics and Natural Science.* Princeton: Princeton University Press.

INDEX

ABOUT THE AUTHORS

MICHAEL POLANYI received his education in Budapest. He was a member of the Kaiser Wilhelm Institute für Physikalische Chemie from 1923 to 1933; in 1933 he became Professor of Physical Chemistry at the University of Manchester, England, and subsequently, from 1948 to 1958, Professor of Social Studies at the same university. He was elected Senior Research Fellow at Merton College, Oxford, in 1959, became an Honorary Foreign Member of the American Academy of Arts and Sciences, Philosophy Section, in 1965, is a Fellow of the Royal Society, a member of the Max Planck Gesellschaft of Germany, and has been awarded honorary degrees at a number of universities. His publications include *Atomic Reactions* (1932), *Full Employment and Free Trade* (1945), *The Logic of Liberty* (1951), *Personal Knowledge* (1958), *The Study of Man* (1959), and *The Tacit Dimension* (1966).

FRED SCHWARTZ received his Ph.D. in psychology in 1959 from the University of Massachusetts. In 1958-1959 he interned at Middletown State Hospital, Connecticut, and then held a post-doctoral research fellowship at The Austen Riggs Center until 1961, continuing as a Research Investigator until 1968. In 1968 he became Head of the Psychology Section in the Division of Psychiatry at Montefiore Hospital and Medical Center. Currently, he holds the rank of Associate Professor of Psychiatry (Psychology) at the Albert Einstein College of Medicine.